T0073893

THE COUCH, THE CLINIC, AND THE SCANNER

ALSO BY DAVID HELLERSTEIN

Nonfiction

Battles of Life and Death

A Family of Doctors

Heal Your Brain

Fiction

Loving Touches

Stone Babies

THE COUCH, THE CLINIC, AND THE SCANNER

Stories from Three Revolutionary Eras of the Mind

DAVID HELLERSTEIN

Columbia University Press

New York

Columbia University Press
Publishers Since 1893
New York Chichester, West Sussex
cup.columbia.edu

Library of Congress Cataloging-in-Publication Data
Names: Hellerstein, David, author.
Title: The couch, the clinic, and the scanner : stories from three revolutionary eras of the mind / David Hellerstein.
Description: New York : Columbia University Press, [2023] | Includes bibliographical references and index.
Identifiers: LCCN 2022044057 (print) | LCCN 2022044058 (ebook) | ISBN 9780231207928 (hardback) | ISBN 9780231557184 (ebook)
Subjects: LCSH: Hellerstein, David—Career in psychiatry. | Psychiatrists—New York (State)—New York—Biography. | Psychiatry—New York (State)—New York—Anecdotes. | Neuropsychiatry—New York (State)—New York—Anecdotes. | Physician and patient—Anecdotes.
Classification: LCC R154.H337 A3 2023 (print) | LCC R154.H337 (ebook) | DDC 616.890092 [B]—dc23/eng/20221117
LC record available at https://lccn.loc.gov/2022044057
LC ebook record available at https://lccn.loc.gov/2022044058

Cover design and illustration: Philip Pascuzzo

Some names and identifying details have been changed to protect the privacy of individuals and institutions.

Contents

PART III. THE SCANNER, 2000–2023

Preface

Who owns the mind? In every age, we humans—self-observing primates that we are—struggle to understand our own mental processes. In every age, we humans dream up a compelling model to capture our deepest thoughts, to explain our motives and yearnings, and to attempt to cure psychic pain.

In the old days of Western religion, under the authority of God, prophets and priests forged this explanatory vision, in which madness could be seen as a gift from heaven, a manifestation of religious ecstasy, a conduit for divine revelation.

Then, just a few centuries ago, philosophers armed with pure logic swept them away: psychosis could be used as a method of inquiry into the nature of reality, revealing new truths about the essence of the world and our existence or, more challengingly, used to throw our usual assumptions and perceptions into disarray, to make it impossible to take our normal world for granted, to amplify Cartesian doubts. For a while, philosophers ruled.

In our postmodern age, though, psychiatrists have pushed priests and philosophers aside, since we—and I speak as a psychiatrist—carry the banner of science. And science, by measurement, observation, and theory, prevails.

Which paradigm of the mind is most dominant? Who cares who owns the mind? *Everyone*, whether we admit it or not. Those who define the mind—and the causes of its perturbations and cures—become arbiters of truth, definers of normality and disorder, dictators of treatment goals and methods, and possibly sources of inspiration.

But here's the thing: though psychiatrists now own the mind, we can't make up our minds *about* the mind. What *is* the mind? A strange persisting state of being that emerges from our hundred-billion-cell organ, connected by 100 trillion synapses, yes, but what is its essence, what are its organizing principles? And especially when its functioning goes awry: what has gone wrong? Our ruling explanation, our entire worldview, our mindset, as it were, keeps changing—one dream, one vision, one paradigm rashly sweeping away the last.

This book is about three revolutionary models of the mind that have emerged, ruled for a brief time, and—one after another—been largely replaced, in a matter of only a few decades. Each model is, in a sense, a vision, a fantasy that organizes the brain's unfathomable complexity in a simplified but compelling way.

But first, let me introduce myself: I am a psychiatrist, trained mostly in New York. I have been on the Columbia University medical faculty since 2000, conducting psychiatric research in mood disorders, supervising residents, and practicing clinical psychiatry. In addition, for two decades I have taught preclinical P&S medical students in the Narrative Medicine Program in a creative nonfiction class, The City of the Hospital, which uses narrative methods to explore the worlds of illness and healing. I have published widely in both psychiatric and literary journals and have written several popular books.

My research has focused on what can be called "clinical therapeutics"—studies of psychotherapy, psychopharmacology, and combined treatments—and has used neuroimaging, MRI, and PET scans to measure the effects of psychiatric treatments. Clinical therapeutics is the amazing and daunting art and science of incorporating the best from research and theory into daily work with patients and determining which treatments work best for which disorder.

Prevailing paradigms and models are necessary to guide such research. They are even more important for treating patients. As a practicing physician it is impossible to hold yourself apart from the moment, to reject all models of body and mind. Instead, you are impelled by the necessity of making decisions with limited knowledge in order to help the person in front of you. This is the essence of medical practice, including the specialty of psychiatry. As a doctor, you thus need a working model, a *vision* to organize the reality that you face daily.

Yet as a practicing psychiatrist it has been dizzying to live through the transitions from one model to the next, in which the ground suddenly shakes and a new system rises up, replacing our previous assumptions and uprooting our complacencies.

Hence my career, like that of many of my colleagues, has been one of continual revolution and upset. Each model of psychiatry has its strengths, its weaknesses and inadequacies, yet each model appears to represent an advance upon previous ones. Or does it? Hence the turmoil and excitement, the chaos and revisioning of what is before our eyes, that each new model brings in. The daily questions, what to do with this patient, how to help this person in front of us who is suffering, flailing, drowning, are answered so differently by each model that it can be hard to know what to do. But we must decide before the end of the hour.

And on a broader level, we wonder: Does each new model actually represent an advance from the previous ones? Does it improve patient outcomes? Does it expand our understanding of the mind and brain in a meaningful way? That is what I hope to explore in this book.

So what are the three visions, the three revolutionary models of mind and brain, that have ruled psychiatry since I began training in the late 1970s?

First came the Age of the Couch. It's hard to imagine today, but in the 1970s psychoanalysis still ruled in America. Its priests were ardent Freudians, psychoanalysts with elegant Park Avenue offices, and they governed via haughty, interminable silences. Their rule began in the early twentieth century and peaked in the 1950s and '60s, but they still held power when I was training in the 1970s and '80s. Every young intellectual yearned to lie on tufted upholstery three or four times a week year after year, unscrolling fantasies before an impassive listener who could free them from early traumas. The mind could be liberated through free association, by regression, and by interpretation of transference and then purged by catharsis.

For decades, this vision was nowhere more vibrant than at the Payne Whitney Clinic in New York City. There, a dozen disciples were admitted to each psychiatric residency class for a four-year period of indoctrination. Those were the ranks I gladly joined in the summer of 1980.

Sure, we were eager to analyze our psychotherapy patients in order to cure their neuroses. But we also yearned to cure ourselves. If we did not enter psychoanalysis on our own volition, each of us was practically compelled to become a patient by our supervisors' earnest advice: personal analysis was the sole way to become a capable psychiatrist. Succumbing eagerly or reluctantly, we found ourselves reclining upon burgundy Ultra Suede or chocolate leather and spent hours tripping royally through our earliest memories, exploring family secrets, our sex- and drug-obsessed adolescences, and the seamy depths of our own minds, hoping to join our professors and psychotherapy supervisors in a state of godlike self-actualization.

But here's the thing: ownership of the mind is never secure. It is constantly in contention, at risk of insurrection. And indeed, the seeds of disruption had been planted even before we settled into our first year of residency training.

In July 1980, when I began residency training, the *Diagnostic and Statistical Manual*, third edition, the *DSM-III*, had already been published by the American Psychiatric Press. With 494 pages and 265 diagnoses, the *DSM-III* quickly replaced the 1968 *DSM-II*, with its 182 diagnoses in a mere 134 pages. The *DSM-III* was shockingly free of psychoanalytic ideas. Thus, said our psychoanalytic supervisors, it was doomed. With its takeout-menu diagnoses, in which you'd choose one symptom from column A, two from column B, and three from column C to come up with your dinner order, it totally ignored the unconscious mind, where all disorder originated. But we gradually became aware that the *DSM-III* was fated to gut our Freudian complacency.

This is why: If you could have a menu for diagnoses, so could you have a menu of treatments to heal these serious disorders—a menu for major depression, for bipolar disorder, for schizophrenia. These new treatment menus, it soon became clear, rarely included the heretofore gold-standard treatment of psychoanalysis. Our analytic supervisors grieved openly, shocked that the latest *DSM* announced the dawning of a new paradigm in which they played almost no role.

The second vision is one that I have struggled to name. Was it the vision of the *DSM* diagnosis? The vision of targeted psychotherapy? Or the vision of the pill—the FDA-approved drug, the sleek capsule, the chemical apotheosis of twentieth-century American pharmaceutical and capitalistic ingenuity that soon began to dominate our work with patients? In a way, it was all three, as played out in the setting of the modern psychiatric clinic. Clearly, not only did the *DSM-III* spawn a host of new diagnoses (all urgently demanding to be treated, of course!), but it also birthed a host of new treatments, both psychotherapies and medications, all neatly sorted by disorder.

And what treatments they were.

The new drugs that we pioneered in the 1980s and '90s in the psychiatric clinic were mostly capsules, not pills. Potent compounds, they were sealed in high-tech multicolored plasticine, ready to deliver their mute cargo of tiny globules optimally to the right part of our patients' digestive systems—like tiny Apollo 13s and Geminis taking astronauts deep into space. Around 1989, inspired by the psychiatrist Peter Kramer, we clinicians—for we took pride in being creatures of the clinic—began resonating to the new SSRIs, the selective serotonin reuptake inhibitors, including Prozac and a half dozen related drugs, which alleviated not only depression but everything from panic attacks to bulimia to body dysmorphic disorder.

The new *DSM-III*-inspired psychotherapies were equally revolutionary. (Suffice it to say, they excluded psychoanalysis). Every month, we scanned the "Green Journal," the *American Journal of Psychiatry*, for the latest updates on the talking cure. There we found rigorous new outcome studies of interpersonal therapy, cognitive therapy, exposure therapy, dialectical behavioral therapy, even streamlined supportive therapy. IPT, CBT, DBT: we threw their initials around, the new "evidence-based therapies." Suddenly there were dozens of scientifically conducted studies showing how targeted therapies could remedy the major *DSM-III* diagnoses, whether depression or panic disorder or schizophrenia.

New psychiatric clinics sprang up everywhere to treat the hordes of newly DSM-diagnosed patients with our snazzy new treatments. The vision "condensed," as my analytic supervisors would put it, around that object: the clinic. The clinic was the holy space where the diagnoses derived from the *DSM*—*and* the neuroactive chemicals developed by Big Pharma *and* the new manualized treatments devised by psychotherapy researchers—could all be put to the test. Only in the clinic could the *DSM*-inspired doctor make precise diagnoses and treat their patients with fancy new interventions. And then—stand back as lives (we hoped) were transformed!

Hence, it is best to call the second vision the Age of the Clinic.

During this dazzling period, I was working in one of the best clinics in New York City, first as a staff psychiatrist, then as assistant director, and then as director, overseeing treatment for thousands of patients, an ever-growing enterprise that eventually had over fifty thousand visits per year. Proudly, we called ourselves "clinicians," swearing allegiance to the *DSM-III* and its ever-plumper updates: the 1987 *DSM-III-R* (revised edition) with 292 diagnoses in 567 pages, the 1994 *DSM-IV* with 410 diagnoses in 886 pages, and then the *DSM-IV-TR*, the weirdly named but barely changed "text revision" of *IV*.

Soon, though, and perhaps inevitably, along with the triple triumphs of our *DSM* diagnoses, our focused therapies, and our potent pharmaceuticals, a nightmare began to emerge. A countervision, an insurrection, a chaotic uprising. Was it creative chaos, from which some vastly improved new order was imputed to gradually emerge? Or was it merely bloodshed?

The clinic's heyday lasted barely a decade. By the mid-1990s, cursed by a first wave of American health-care reform, we clinicians faced crushing rules dictated by newly powerful health insurance companies, which began a merciless process of "managing" care.

It was war. A massive collision between Wall Street's desire for maximal economic return and the clinic's lofty ideals—in which patients could be given enough time and attention to get a proper diagnosis and receive treatment from expert doctors and therapists until they began to recover. Instead of a therapeutic paradise, we were suffocated by the number crunchers, and the clinic soon became a sweatshop. Overwhelmed by massive caseloads, our therapists began shortening sessions, seeing patients by the group or every quarter hour, squeezing more bodies into each eight-hour day. We psychiatrists saw patients once per month, then every two or three months for "med checks" to renew their Celexa or Risperdal supplies—for fifteen, then ten, then five minutes, until finally we were basically waving a prescription pad in front of a crowded waiting room. Goodbye analytic couch; goodbye luxurious endless emotionally charged silences. And hello fluttering prescription pad and drive-through psychotherapy!

And then, in the early years of the twenty-first century, my third decade of practice, yet another vision began to emerge. We abruptly found ourselves living in the era of the whole human genome, of "personalized," or what was soon renamed "precision," medicine and the visualization of "task-based" brain-center activations by a dazzling array of neuroimaging techniques using MRI and PET scanning machines. Not only that, we were clearly at the first stages of trying to tweak our brains' "epigenetics" and "retuning" aberrant brain circuits by direct-current electricity or profoundly disruptive drug infusions. We are now, in short, immersed in a neuroscience revolution in psychiatry.

Is this third era the "Age of Neuroscience"? Willy-nilly, since the year 2000, I've joined a whole corps of science-obsessed psychiatrists, which is just now coming into its own as "clinician-neuroscientists", as "circuit psychiatrists." However poorly prepared we may be to understand or incorporate advances from neuroscience, we do our best to progress into the new era. We are forever scrutinizing the research literature, reevaluating our methods, squinting yet again at our patients: we seek to tunnel our way into a truly scientific psychiatry. True, our two most powerful instruments, the MRI and PET, are both "scanners"—but so are *we*.

Hence, I've settled on calling our emerging new vision the "Age of the Scanner."

The physical soul of this era is, as I've said, embodied in our brain-imaging machines. The magnetic resonance imaging (MRI) scanner cleverly uses computers to generate images that show not only the structure of the brain but also its connectivity, and, more amazingly, it lets us see neural circuits at play as blood surges to particular hubs. Our positron emission tomography (PET) scanners use radioactive tracers to measure chemical levels deep inside our subjects' brains without probes or needles or neurosurgical instruments and can thereby test the effects of disorders and their treatments on levels of chemicals in the brain.

To me, the MRI is the more revolutionary of the two technologies. Using fluctuations in magnetic signals, the MRI can capture the way that *thoughts*, moment by moment, alter brain activity. It can demonstrate, how, over time, our patients' customary behaviors sculpt the brain's insulated white-matter circuits and can resize and reshape their very brain cells. In the old days, our patients lay upon an overstuffed couch, letting their minds wander, awaiting their psychoanalyst's wise interpretation. Today, our research patients lie stretched upon a scanner bed, peering through goggles at scenes of virtual reality, instructed to cogitate and problem solve, or told to close their eyes and summon feelings of sadness or euphoria or simply let their minds wander.

We psychiatrists-turned-neuroscientists have begun to find specific brain areas that are activated when our patients learn to juggle or when they wrestle with moral dilemmas or when they are stirred by sadness or anxiety. Psychiatric journals are now chock-a-block with gene heatmaps, colorful brain images, and complex statistics. The Age of the Scanner has brought psychiatry more fully into the world of medicine, ever closer to finding pathological anatomy and disease-associated genes.

Already, the Age of the Scanner has begun to deliver its first practical treatment advances. We are now starting to trace the brain circuits that underlie the *DSM* psychiatric disorders and to devise ways to modify their activity. We intend these new drugs, such as ketamine, and brain-stimulation devices, such as the transcranial magnetic stimulator, to rapidly reset these aberrant circuits. We prescribe ancient approaches like mindfulness meditation and yoga, which have been shown to have a profound capacity to retune these circuits. Furthermore, neuroscience inspires us to take new looks at existing medications, whether FDA-approved drugs, drugs of abuse, or plant-based compounds that have been used for millennia in indigenous spiritual experiences, and to employ them for brain-circuit retuning. We continually translate our latest findings from the lab to patient care and back to the lab again.

Already this latest vision has become the most compelling one of all.

But one more thing, oddly enough, connects the Age of the Couch and the Age of the Scanner.

The story.

The psychoanalyst and the behavioral neuroscientist are strange bedfellows in their appreciation of the value of stories. Partly, this has to do with the never-ending talk of psychotherapy, the healing dialogue between doctors and patients, which, once started, rarely terminates. For the psychoanalyst, dreams provide direct entry to the unconscious. Interpreting dreams and fantasies—recasting and retelling these unconsciously derived tales—was the method by which patients sought cure.

Surprisingly perhaps, stories are *also* central to neuroscience-based research in the twenty-first century. Today, in the Age of the Scanner, neuroscientists speak of "autobiographical narrative" as a way of *organizing* the brain. Stories (whether heard or told) allow our brains to connect and make sense of wildly diverse experiences. To the neuroscientist, stories propel the brain's continual and unending process of restructuring and reconnecting itself.

To make sense of changes in the psychiatric profession over the past several decades, to explore these three visions, it is perhaps inevitable that this book should be a collection of stories. Both neuroscientists and psychoanalysts prize *autobiographical* narrative for its organizing power for mind and brain.

These fourteen tales are indeed autobiographical narratives, written from my perspective as a psychiatrist practicing in New York City from the early 1980s to the present day, a physician who spends his time working with patients, doing research, and helping run clinics and hospitals. More precisely, I see them as adventure stories. The book documents my own internal battles as explorations of the self, full of wonder and angst and occasional surprising revelation, as well as my vicissitudes in what the poet and physician William Carlos Williams called "the city of the hospital": amid labor strife, cutbacks, and mutating bureaucracies, my fellow doctors and I find amazing science and unexpected poetry in hallways and waiting rooms, encounter brilliant and not-so-brilliant students, and have to reckon with our own limitations and frailties. And of course, my daily adventures center on the innumerable patients a doctor encounters every day, sometimes baffling and frustrating and often inspiring.

My approach here differs from other recent and classic books about psychiatry. Thomas Szasz's 1961 classic *Myths of Mental Illness* and David Heath's more recent *Pharmageddon* are compelling polemics that add little to daily clinical practice. Jeffrey Lieberman and Ogi Ogas's 2015 *Shrinks: The Untold History of Psychiatry* is a broad history of the field that induces useful humility in present-day enthusiasts. Alan Frances's 2013 *Saving Normal* is a well-deserved critique of *DSM* overreach that can temper a doctor's diagnostic enthusiasms. This book is not intended as a scathing exposé; instead, it is an effort to explore the lived experience of practice—by a doctor who is *in* practice—over decades of ever-changing theoretical models.

My writing has been strongly influenced by literary physician-writers, including the surgeon Richard Selzer, the immunologist Lewis Thomas, the infectious disease specialist Abraham Verghese, and especially the poet-physician William Carlos Williams, whose *Doctor Stories* features an engaging and highly imperfect protagonist. *Letter to a Young Female Physician*, by the Massachusetts General Hospital internist Suzanne Koven, continues this medical literary tradition, with an increasingly urgent goal of rehumanizing medical care. Rehumanization of care is crucial for psychiatry, where many practitioners have abandoned psychotherapy and become mere pill pushers and many people struggle to find accessible, affordable, and humane treatment.

The fourteen tales of in this book span the three-plus decades since I entered residency training and began to work as a practicing psychiatrist and then as a psychiatric researcher. The organization is thematic and roughly chronological, from the Age of the Couch, framing my experiences from 1980 to 1994; to the Age of the Clinic, from 1985 to 2000; and the Age of the Scanner, from the late 1990s to today. Ages, like stories, often overlap, since it is rare for one age to end abruptly before the next begins. Instead, they often coexist for a time, reflecting natural reality. My goal in telling stories is to engage deeply and shed light on serious issues within the recent history of psychiatry, writing from the crucial perspective of the actual implementation of care using three vastly different theoretical models.

In telling these stories, I realize that each age has remade *me*. I began as a psychoanalyst in the making; I then became an enthusiastic clinician, a devotee of the *DSM* method for diagnosis and designing new treatments; and now, in recent years, I have been remade yet again as a circuit psychiatrist, exploring neuroscience's tools as they begin to be applied to the consulting room and now, very suddenly, to the unexpected new world of psychedelic research.

As such, these stories are efforts to explore the different visions of these three revolutionary eras of the mind. Each of them, I believe, describes a voyage in foreign territory, an investigation into something unknown, and an amazing adventure in which my colleagues and I—and our patients and research subjects—all seek to define the bounds of new continents.

THE COUCH, THE CLINIC, AND THE SCANNER

PART I

The Couch, 1980–1994

CHAPTER 1

The Work

Learning to Do Psychoanalytic Psychotherapy, 1980–1984

"Well?" she says.

"Well what?"

"Do you want to or not?"

I look down at the floor, at the walls, out the window through the haze of cigarette smoke—everywhere but at her.

Ms. J. My twice-a-week psychotherapy case, an art school dropout who worked at a dead-end job at an insurance company, is my first psychodynamic case. I'm supposed to engage her in depth therapy, exploring her fantasies and dreams, her darkest urges. Today despite the chilly weather she is wearing a thin summery dress. Her legs are bare. She keeps crossing and uncrossing them, licking her chapped lips. She pushes her sandy hair behind her ear; it falls back over her face, and she pushes it back again.

She had begun today's eight a.m. session as usual, recounting the events of her weekend. We've had a sort of flirty banter for the start of therapy sessions, our usual chat.

"What do you think I did this weekend?" she asks.

"OK, so what did you do?" I respond.

"What did *you* do?" from her. "Oh. I can't ask, you're not allowed to tell." Then the tone changes. She starts telling the details of a dream from Saturday night—about a man she met on a bridge. She was crossing the Brooklyn Bridge, but then she was naked, and she was in his apartment, she wasn't sure how she'd gotten there or who he was, but she was somehow wildly excited.

I wait quietly, ready to sift through her associations: How did you get to his apartment? How did it feel to be naked? What do you mean, wildly excited?

Until this unexpected moment—this outburst—this asking if I want to . . . and I realize, it's hardly about the dream, it's a question she's asking me. A proposition. Do I want to? Until this moment, I had believed things were going reasonably well.

She turns away and begins to weep.

"I'm so humiliated," she says. Her face is flushed, her ears bright red. "You don't want me, I should have known! People are always rejecting me—I'm hideous, this just proves it."

"Hideous? No! . . . what do you mean?" is all I can manage.

My heart pounding, I try to focus, to maintain the dialogue. Do I want to, how does she mean "want to"? Should I ask that? It seems ridiculous. I need to say—to ask—something. So, I ask, can she tell me more? *Always ask questions*, they say, *keep the conversation going.*

Abruptly, Ms. J stands, grabs her coat, and flees, slamming the door behind her. A moment later, the door opens. She comes back for her purse and her art portfolio and slams the door again.

I sit there, stunned. A lipstick-stained cigarette still glows in the black Bakelite ashtray on the edge of my desk.

This is the third year of my psychiatry residency. It's November 1982; I am twenty-nine years old and have been in school since the age of four. After five years in California, including a dropout year and four years of medical school, I've been shocked by coming to the East Coast. During my wander year I'd hitchhiked down the California coast, worked as a bicycle messenger in San Francisco, in a restaurant/bookie joint, and as a door-to-door pollster, and wrote endlessly, determined to become a writer. Then I finished college and started medical school at Stanford, living in the laid-back sunny flatlands of Palo Alto, learning about serotonin and revolutionary Hodgkin's disease treatments and doing psychiatry rotations at the Palo Alto VA Hospital where Ken Kesey had lately worked as an orderly, where he was inspired to create Nurse Ratchett. In contrast, New York is utterly overwhelming: a culture shock, to put it mildly. Nothing is laid back here. I had survived my internship year mostly on medical floors of New York Hospital. Then, six months ago, I moved from doing inpatient work with psychotic people to the outpatient clinic, to work with what we still quaintly think of as "neurotics."

I entered residency training with a strong belief in the scientific method, in the primacy of observable facts—spots on X-rays, abnormal blood test results, tumors you can palpate by hand, which had been boosted by my time at the high-tech glitter palace of Stanford medicine. Now, after a few months of learning psychotherapy here in New York—an old-school and rigorous type of therapy—it seems to me there's no such thing. Everything is slippery and subjective. I'm adrift in a sea of emotions. Not just familiar emotions, expressed directly, as in "I'm kind of sad today," but a turbulent, encompassing theater of feeling, of murderous rage, vengeful jealousy, endless wounded silences, sudden unfair accusations—and now, somehow, my patient's heartfelt expression of desire.

At Manhattan's Payne Whitney Clinic, the events of Sigmund Freud's life remain vivid, as though he had conceived of the unconscious mind only a few years back. The interpretation of dreams, the ego and its mechanisms of defense, the mechanistic model of early psychodynamic theory, the later, more sophisticated, object relations theories—these are dominant, unquestioned principles in this venerable building, set on a bluff overlooking the East River. I had looked into the various psychiatric training programs when I was just about to finish medical school in California, and after much deliberation settled on this place, one of the great temples of psychiatry. Of *psychodynamic* psychiatry—a form of depth psychology whose primary focus is to reveal the unconscious contents of the patient's psyche, to resolve conflict, to work through something or other. Psychodynamic—anyhow, we are told here, it's the only kind that matters!

After an internship year mostly spent on medical floors, my classmates and I, a dozen in all, moved onto the psychiatric hospital for a year of inpatient work that in many ways resembled our medical rotations, except that we treated people with schizophrenia and depression rather than heart failure and diabetes. Now, in the third year, working at the outpatient clinic, we are charged with mastering the dark arts of Freudian psychotherapy. Difficult issues routinely emerge here, painful feelings, I knew that. As a college student, a history and literature major, I found Freud's ideas intriguing, linking the worlds of imagination with those of science and psychology. Now that I am charged with applying them, I am finding them impossible.

I need to leave my office. Not that I have anywhere to go, but just to escape this hazy cubicle, echoing with traffic from the FDR Drive. I flee to the lobby, where I take comfort amid the other residents in long white coats, shuffling across Payne Whitney's marble floors past the battered wooden paneling of its corridors.

Ducking into the mailroom, I extract a thick stack of psychoanalytic reprints from my box—Freud and Rank, Kohut and Kernberg: assignments for the coming week. Our readings on borderline personality. A relatively new concept in psychodynamic psychiatry, borderline is a uniquely unstable type of personality, first described as "borderline schizophrenia" but turning out to have nothing to do with psychosis. One of our supervisors describes borderline as "a stable form of instability," a state of continual crisis, involving behaviors like self-cutting and pill overdoses. Psychologically it involves a lack of stable "self-experiences" and consequently a continuous difficulty of living in the world.

It's reassuring to hold these still-warm photocopies to my chest as I walk out into the chill Manhattan winter air, my heart still thumping. Luckily, there is no sign of my patient as I approach the avenue. Maybe *she* has a touch of the borderline? I have seen scratches on her wrist and old, well-healed scars dating from her college years. What will I tell my supervisor, Dr. Banks, when we meet this afternoon?

Somehow, I realize as I stand on line at the deli across from the hospital waiting for coffee, somehow for this type of therapy to work you're supposed to use your own personality to interact with the patient—to use your "self" to cure them in a small, smoky, not quite soundproofed room. With silence, with probing questions, letting the patient regress and develop strong feelings toward you—transference—then making interpretations to move things along, to heal their psyche. But exactly how are you supposed to do all that?

"Hey buddy!" My coffee is ready. Clearly, in these ambiguous silences, I am failing utterly. I throw down 40 cents, grab the blue-and-white paper cup—WE ARE HAPPY TO SERVE YOU—and head back to my office.

"I must admit . . .," I say to Dr. Banks, during our supervision session later that day. We are in my office, and I'm beginning to talk about this session with Ms. J. "I must admit. . . ."

I am trying to find a way to say that for the past several weeks with my patient something has been growing in the room. Something impalpable, impossible to name. But nevertheless intense—feelings, thoughts, fantasies I'd rather not have.

Dr. Banks, my supervisor, sits in the chair where Ms. J had been a few hours ago. A stylishly dressed psychoanalyst with unruly hennaed hair and spike heels, she has no discernible sense of humor. Today, as always, she smokes a thin brown cigarette, something foreign, practically a cigar. I'm embarrassed to see that the stub of Ms. J's cigarette is still in the ashtray, marked with a gash of red lipstick.

Dr. Banks flicks an ash. She is saying something about "transference," putting the accent on the second syllable—trans*fer*ence. "Ms. J's trans*fer*ence."

I know what *trans*ference is—it's the patient's projection of feelings toward important "objects" (i.e., people) from their past life onto the therapist, an emotional response that was once appropriate but now outdated, out of sync with reality.

I avoid Dr. Banks's gaze, and my mind wanders away from the room, as it so often does these days. I know that the ability to work with transference, however it's pronounced, is essential to this type of psychotherapy ("the work," as Dr. Banks puts it) because you are thereby grappling directly with the unconscious, like Jacob wrestling with a God-sent angel. It's from Genesis—I recall that story from Hebrew school—about a battle that continues through the night, with neither Jacob nor the angel emerging victorious. But Jacob ends up being touched by God, forever changed. Which maybe also describes psychodynamic psychotherapy? But no, this can't be a religious experience, or is it?

Regardless, there's no avoidance here, no abstraction, no glossing over; it's all so intensely *here and now*, replayed in the sanctity of the consultation room. Even the boredom is unbearably intense! This *is* therapy, we learn, the only true cure. The rest is barely chatter.

And yet I am an utter failure at "the work"!

That's one way of looking at things, the analytic lens, so to speak. But often enough, for whatever reason, my mind will flip and reject it all as nonsense. At those moments, this tension, this trans*fer*ence stuff, will seem ridiculous, a distraction from the important real-life issues my patient and I are discussing: issues in her relationship with her on-and-off boyfriend, an alcoholic bartender; her teenage anorexia and half-hearted suicide attempts; her fights with her dysfunctional family up in New England; the panic attacks that wake her in the middle of the night; her struggles to pay the rent, to find affordable studio space, to make connections in the art world while she tries to make it as a painter. Not to mention her drunken downtown wanderings, the clubs where she meets guys for random sex, sometimes not sure how she got home, other times waking up somewhere in Alphabet City. All of which she swears never to repeat.

Transference, shmansference! I just want to get past it. So, I've tried to be extra-nice to my patient, pleasant, friendly, maybe even flirting a bit. Why not? I think. She's having such a hard time in life, why not make her time here as comfortable as possible?

"I'm not exactly sure what's happening," I tell Dr. Banks. "I thought things were going pretty well. She's really engaged. Recently she's been talking very openly for the first time about all kinds of personal things—dreams, sexual issues, relationships. So, this is completely out of the blue."

As I'm talking, I recall the level of detail Ms. J had offered lately about her sex life and the physicality of her presence in the room. I mention to Dr. Banks the dream she had described earlier in today's session and recall the smoke lifting from her Pall Mall. Speaking of cigarettes, when I leave here at the end of the day, I always reek of smoke—sharp cigarette smoke, the musk of cigar smoke, hints of marijuana. In my hair, in my clothes, you can't shower or dry clean it away. My wife objects: can't you tell them not to smoke? Tell them you're allergic! But I can't—the fumes themselves seem essential to the treatment.

"*Doctor*," exclaims Dr. Banks. "Do you think there may be something you are avoiding?"

Suddenly, sweat beads on my upper lip. My ears start ringing, my mind is chaos. Maybe it is less like Jacob with the angel, more like Odysseus bound to the mast, struggling to resist the songs of the Sirens.

What can I possibly say? That I'm in love with Ms. J?

Luckily Dr. Banks takes mercy on me. "You had no sense there was an erotic element to the trans*fer*ence?"

"I guess . . ." I take a deep breath, and the ringing begins to fade away. "I guess I should have been aware of it earlier." I think for a moment, debate whether to go on. It's so personal. Way too personal. Somehow . . . seamy. But it's essential to the treatment approach. I can't hold back. "You know, she did miss two appointments after she ran into me at the movies with my wife. But . . . I didn't really make much of that."

I flip through the spiral-bound notebook where I keep "process notes"—phrases red underlined in places, my barely decipherable handwriting—through scrawlings of the past few weeks. I suddenly recall how upset Ms. J seemed after I came back from my honeymoon, with a wedding ring and a tan. Do you have any thoughts about me being gone? About the ring on my finger? She had laughed. I'll let you know if I do. And she hadn't answered—until now.

"And what she said about my wedding ring, too."

Dr. Banks nods. "It seems clear," she says. "Not only is there an erotic element here but also a sense of competitiveness and jealousy that you need to pay more attention to. In the dream material, of course, and her mocking you, and in mocking you about your wife. Your patient seems to feel she can't have the success you have—as a result she wants *you*."

"So . . ." I try to catch my breath. The man on the bridge. The apartment. She was naked in her dream. She wanted to be there, she said, but wasn't sure how she got there. So was it about her downtown drunken wanderings? Or maybe was it about therapy, about me? But why me?

"First of all, don't worry," says Dr. Banks, going into teaching mode. She is saying something about how with male therapists the "transference neurosis" often focuses around the wish for approval, reproducing power relationships in contemporary society.

But is that true with me and Ms. J? OK, I'm her therapist, but even with her lousy day job she makes more money than me, and she lives in her own apartment too, not some beaten-up hospital-owned linoleum-floored quasi dorm where my wife and I live. Plus, she's also an artist: her work is good.

I just don't see it with Ms. J.

"Often this can be an eroticized trans*fer*ence, as you see with your patient," Dr. Banks continues. "Some radical feminists feel that, as a result, women should only go to female therapists, but I do not entirely agree." She does think it's important to be aware of this tendency, though, and to address it early, because otherwise it can get out of hand.

"OK, so what should I do?"

"Call your patient," says Dr. Banks irritably. "Get her to come back for discussion. OK?"

"Fine," I say, though I would be thrilled to never see her again. Or Dr. Banks, for that matter. Somehow her using the word "entirely" has hurt my feelings. The edge to her voice sounds personal—maybe *some* men can get it, but not *you*, Dr. Hellerstein! Maybe women are more intuitive, better at this? So, does this mean that men shouldn't be therapists at all?

Dr. Banks stands to leave, and I have survived another supervision.

I phone up Ms. J the next day.

It does not go well, to put it mildly. "You really think I want to come back and see *you*?" she says. "I'm sick of this whole thing, of what men do!"

At this point, it would be fine with me to end it, save for the prospect of facing Dr. Banks's disapproval: *You coward! Failure! You'll never do "the work"!*

"Well," I say, "You . . . you've been coming for so many months. You've said you really felt comfortable, that it was helping. Don't you think it's worth coming in just once more to figure out what's going on?"

A long silence. I think I hear her sighing.

"Sorry, did you say something?"

"OK," she says, "OK."

The following week Ms. J appears at my office twenty minutes late for her appointment. For some reason today she is wearing her painting clothes—paint-spattered shoes and jeans, a loose flannel shirt, a ragged army coat. There's even paint in her hair. The room is icy—a frigid January wind blows off the East River. Keeping her coat on, sitting on the edge of her battered wooden chair, she berates me.

"I still don't understand," I say.

Ms J stares at me icily. She snaps back: "Of course you don't. Just like my . . ."

"Wait a minute," I say, pulling my white coat close over my chest. "I mean specifically. You said it's something I *do* that's the cause of the problem. Could you tell me what it is?"

An infinite pause, and then she tells me. "I don't know how else to put it," she says. "You *flirt*. You don't want me to be angry with you, you don't want me just to like you, you want me to be in love with you. *Why do you do that?*"

Silence.

"I . . . I'm not . . . I can't say I'm aware of doing that." Actually, I can't honestly say that I'm entirely unaware of doing that. But I can't say that to her, not in *her* therapy. I can't say what I want to say; I'm saying the opposite of what I mean. And I have absolutely no idea why I do it.

God! It's ridiculous! It's as though *she's* the therapist and *I'm* the patient. At this moment I'd do anything to return to the previous balance. She's struck a nerve, too. In fact, this isn't the first time this has happened. A year earlier, with my first once-a-week psychotherapy case, a gay man, things went well for a while, then suddenly there was an inexplicable rash of missed sessions. And one day *that* patient said something mysterious about realizing I didn't feel the same way about him as he felt about me, so he was quitting.

My patient is watching me. Her hands are on her thighs, she is leaning forward, her mouth open. The way she's sitting I can see down her shirt: why doesn't she wear a bra so I don't have to see her damn pink nipple? Which I really can't unsee. This is nuts! I just really want to jump her bones, and I sense that she . . . she feels the same way. Why would I even be thinking this sick way?

"Well, how does that affect you?" I say at last.

"How it affects *me* isn't the point," she says.

She stands; she's going to have to think about whether to come back or not. She buttons up her coat, avoiding my gaze. Then she's gone, this time leaving the door wide open.

"There's . . ." I admit to Dr. Banks at our next supervision session a week later, "there's, um, an element of truth to it."

Today I've made sure to clean out the ashtray.

"A countertrans*fer*ence issue?"

Transference, it turns out, is only one half of the therapeutic relationship. As transference describes the patient's feelings for the therapist, so *counter*transference reflects the therapist's feelings toward the patient.

"It could reflect unresolved issues from the therapist's own life or give clues to issues that the patient is struggling with. Do you think that's relevant here?"

"Maybe . . ." I say, then sink into silence.

"Either way, understanding these is crucial for progression in the work."

Unresolved issues? These days, besides my entirely unwelcome feelings for Ms. J, my "ego-dystonic" feelings, as my supervisor would put it, I am painfully aware of all the unresolved issues in my life. Going to medical school, becoming an MD, choosing psychiatry. Getting married. With each step, I have been overwhelmed by doubts and unsure of the choices I am making willy-nilly one after the next, feeling pushed forward by the tide of life, wanting more, wanting everything, not to have to choose even something as basic as whether to be a writer or doctor. I cannot make up my mind; I want to be both. And then, in the thick of things, deciding to become a psychiatrist. I am here, yes, but is it somehow a matter of convenience rather than a true decision?

Now, in the frenzy of learning psychodynamic therapy, it's like a ripping away of bandages, having no choice but to jump into the morass. It's scary as shit! Hence, my uncertainty about whether I can do "the work," regardless of whether I "believe in" it (a whole other issue).

Anyhow, who's training whom here? To a great degree it feels like Ms. J is training *me*. Or trying to, since I don't seem to be a very quick learner. This is the great mystery of training programs, medical or psychiatric, how raw young doctors are thrown into situations they can't possibly be prepared for and somehow come out the other end as well-trained professionals, order emerging from chaos.

Dr. Banks is watching me with what looks like amusement.

I gather my thoughts like trophies. "I . . . I think so. It's as though I want to influence how women feel about me. Sometimes I'd rather have the feeling of an infatuation, I guess, than the reality of everyday feelings."

"A state—*of love*?" hazards Dr. Banks.

The words seem both grossly inadequate for what I feel toward Ms. J and at the same time painfully exact. Yet . . . is it love? Or just desire? Either way, it is like an infection, a fever, a virus, the way my mind keeps returning to her voice, her face, her body. The problem being that for a time I almost liked that feeling, only not when it blindsided me with its intensity.

"I think—it's more like being halfway in love. Not enough that you'd *do* anything about it—but a pleasant preoccupation." It sounds idiotic now, but at first it had seemed like something admirable, inherently good. "Maybe . . . maybe, to take our minds off more uncomfortable things."

"Like what?"

I think for a while. "She's so enraged," I say. "I'll say something, and she's totally furious. I'm, like, why, what did *I* do? Then . . ." I can't figure what to say. "It's not just the anger. She's also . . . she's very needy and always wants to be reassured, but then she's angry if I try to reassure her. And . . ."

"And?"

"I don't know, and, she says she isn't, but I think she has these incredible feelings of jealousy." A long pause. "And maybe something about *my* feelings."

"Your feelings?"

"If I'm uncomfortable about being really close to women, really intimate in a nonsexual way . . . maybe it's easier to have things sexually charged. And most of the time, with friends for instance, that works reasonably well. But . . ."

"But in a therapy situation?"

"I guess it leads to disaster."

Dr. Banks almost smiles. "So," she says. "Do you think your patient will come back?"

"Oh, sure," I say. "She hasn't finished telling me off."

When I go to our weekly third-year resident process group the next afternoon, I look around the room with fresh eyes. Can I be the only one of us going through this kind of thing with psychotherapy patients?

I can only guess what my eleven residency classmates face in the intimacy of the consultation room. Georgie, for one: he is terminally vague and befuddled, always fifteen minutes late for every meeting. How does he do with his patients? And what about Zooey, who scrubs his face incessantly and whose deeply bitten nails seem to ooze angst from his very core? I can't imagine him being particularly capable with countertransference issues. Annabelle, a Manhattan native, languorous and aristocratic, who gazes out at all of us benighted peasants with an uncanny world-weariness and who is now six months pregnant—you'd think she would be good at this, but for some reason she hasn't been able to get her patients to discuss their feelings about the baby inside her. And what about our triathlete, McNeil, my best buddy, with a well-deserved reputation of being a serial dater-and-dropper: supposedly he has trouble keeping his patients in therapy past a few weeks, and we all speculate how that might relate to his romantic life.

All of us, the dozen in this room, are smart and accomplished, graduates of the best colleges and medical schools. We are the best hope of psychoanalytic psychiatry, which, truth be told, has seen better days. Twenty years ago, analytic therapy was the highest goal. Young psychiatrists fought to get admitted to psychoanalytic institutes. If you were turned down for analysis, deemed to be "not a good analytic candidate," you would be crushed. Not so now. Now the focus of psychiatry is turning to new models; everywhere you look someone is debunking Freudianism as a relic of the past. Now, the various institutes compete to get candidates. We're the best of this season, that's about it.

Every week, we gossip and complain and jostle for position (who will be anointed chief resident? Surely not me!); we snipe about one another and our professors; we form alliances and engage in turf wars. And on the way, overseen by our nearly mute, suit-vested, bewhiskered, pipe-sucking psychoanalyst group facilitator Dr. Thad, we learn a bit about "group process."

Today, the topic is love.

Georgie, coming late to the group as usual, interrupts Annabelle to blurt out that his patient, a middle-aged woman, a mother of three, has fallen in love with him. She pages him through the hospital operator; she waits outside at the end of the day; she even found his home address!

Dr. Thad raises a hand. Has anyone else experienced something similar?

Well, Annabelle was stalked by a manic patient from inpatient. McNeil, our romantic player, has been propositioned by two young male patients. Even Zooey has been the target of love, from a demented lady on the geriatric ward.

Cupping his unlit pipe, Dr. Thad confides how one famous analyst came back to his apartment building to find a patient in his very bed, a young woman who had convinced the doorman that she was his sister. We have met this famous analyst, potbellied, slope-shouldered, bald. Truly, transference must be blind!

Then I can't help blurting out: "Well, what if a psychiatrist did fall in love with a patient, what if it was real, if it wasn't countertransference?"

Everyone stares at me.

"Just theoretically," I add.

"What do people think?" asks Dr. Thad. But he immediately starts telling of famous analysts who married their patients in the early days of psychoanalysis, before the development of ethical guidelines that forbade such behavior. As he tells it, it's an issue of escaped countertransference, an inappropriate enactment by the therapist, essentially a *folie á deux*. "Probably representing a psychotic break by the analyst."

Time is short, and everyone starts to gather their photocopies, their backpacks, to extinguish their cigarettes, ready to head back to our patients.

"This is important," Thad says as we walk out. "We'll come back to it next week."

Annabelle catches up to me in the long hallway. "You know, at this other program in town," she says, "there was this resident who met up with a patient at a club after work. The administration found about it and threw him out of the program." Annabelle's office is just down the hall from mine, and she gives me a look as she opens her door.

In God's name, what kind of work is this? Can we possibly muddle through our own psychopathologies to help our patients with theirs?

Ms. J does come back, and for the next month all I hear is rage. I try to do things differently. I try to do nothing, to be silent. I resolve not to flirt. I listen longer; I ask more questions: Is that all there is to it? What does that make you feel? What does that bring to mind? Anything else?

Dr. Banks tells me to point out paradoxes, inconsistencies, but trying that mostly just elicits fury.

Ms. J is furious at the men who've tried to seduce her (or in some cases, the men uninterested in seduction, including her previous boss); she's furious at herself for wanting men's approval and for the panic that overwhelms her when she feels rejected. The art professors, more than one, who praised her work, who invited her to their studios or who came to her studio to see art but expecting sex, one who nearly raped her. Once, on her initiative, she gave in to a famous artist, a disgusting man, because she thought it might help her career. More specifically, she's furious at me. I've made her avoid the central issues. I don't really want to help her, she says, I just want to be admired. She's always disagreeing, arguing, confronting me, mocking my interpretations.

She laughs, no matter what I try. So why on earth does she keep coming?

Over time I learn more about Ms. J's backstory: her mother's death when she was eight, her father's alcoholism, his unending and futile struggle to care for her and her two sisters in small-town Maine. A whole world opens as I listen to her for hour after hour: How she got to New York City on an art scholarship but lost it after two and a half years. Her subsequent depression, her self-cutting, her wild nights at bars downtown seeking consolation by picking up strangers, her mentorship from a gallery owner who encouraged her to work on larger and larger canvases and who assaulted her one night, her drinking to the point of oblivion with her bartender boyfriend. And how she gradually settled down, started therapy, getting a regular job, as lousy as it might be, that covers her rent. *This phase of your life counts as settled down?* I think, but say nothing. She still occupies my thoughts when I am away from Payne Whitney, which worries me, but over time I think about her less. I don't know if that is good or bad, but as winter turns to spring, she is not on my mind much except when she comes to my office.

And I keep trying to make sense of her. I keep paging through my xeroxes of the analyst Otto Kernberg's classic *Borderline Conditions and Pathological Narcissism*. I read about "oral rage" and "sadistic precursors of the superego," about "structural derivatives of object relations." It's rough going, but I think I get it. According to Kernberg, people with borderline personality *organization*—which is found among people with narcissism and antisocial personalities, as well as those with borderline personality disorder—can't integrate their positive and negative experiences. They rely on "splitting" as their primary defense mechanism, seeing others as either all good or all bad or alternating between all good and all bad. In their lives this leads to catastrophe. Since they don't have a stable self, they can't imagine others as having stable selves either. As result of such "ego weakness" they lack anxiety tolerance and make all sorts of impulsive acts to try to feel better—sexual acts, self-injury, drug use.

In Kernberg's text I see images of Ms. J's self-injury, her self-hatred and breathless rages, but almost nothing of her as a person, an individual. She does act this way, but is it really her "personality," something fixed? Or perhaps a state, a condition, something she can work through in therapy? By the way, in working with such patients, one catastrophic mistake is for the therapist to be uncritically "nice."

"Have we perhaps heard enough of her anger?" Dr. Banks asks one day in supervision.

Finally, I am ready to confront her back. "Have we perhaps heard enough of your anger?" I echo to Ms. J the next morning. A cloudy April day, wind blowing clouds over the East River, her clothes splattered with black paint.

She stares at me, for once at a loss for words.

"What I mean. . . . What I mean is, isn't it time to stop just being angry, and . . . maybe to start trying to figure how you can deal with it better, so it doesn't keep ruling your life?"

She actually smiles at me, a friendly, nonseductive smile. "Maybe," she concedes.

And we start seeing what Dr. Banks describes as "a progression in the work."

When I finish treating Ms. J a year later, in the spring of 1984, when I am completing residency training, her life is going pretty well. She has broken up with the nasty barkeep. She finds shared studio space in Long Island City in an old cork warehouse. She has begun dating a guy who works a few blocks from her insurance company. They buy a used car to go upstate. She lands a new part-time job and applies for a scholarship to get a master's degree in painting. She even has some of her work shown in a few group shows and gets written up in a magazine; a photo shows her standing in front of one of her huge splattered canvases, grasping huge paintbrushes.

She sits before me now, holding an unlit cigarette, thinking of something.

"What?" I say.

"I was just realizing, I'm so over you," she tells me. "That's good, right?"

I nod. "I think so." I'm kind of over her too, also in a good way.

In sessions, the storms of rage have largely faded. I've stopped thinking of her as borderline, despite her history of wrist cutting and all her other diagnosis-confirming behaviors. The transference is actually fairly pleasant—or maybe it's not transference, maybe it's just our relationship, the way Ms. J and I are working together. And—though I wouldn't tell Dr. Banks—I've more or less stopped obsessing about her when out of the office. Except when she appears in my dreams; I mean, who can control their dreams?

A reasonably good resolution so far, we'd all agree.

Ms. J contends that I've gotten better too—she'll joke about how hard a job it's been for her to break me in as a psychotherapist: what would I have done without her?

And she's got a point. It's hard to say exactly how I've changed, but I have. A year ago, I didn't know the first thing about all this therapy stuff, which seemed somehow so undoctorly, so unscientific, so unamenable to any type of objective measurement. However painfully, however awkwardly, she has helped me take a few steps toward becoming a psychoanalytic psychotherapist. I'm hardly there yet; I'm just not entirely incompetent anymore.

Just a few years back, I went from being a medical student to becoming a doctor. Now I have weathered a personal crisis (thanks to my own therapy) and have made the first steps toward yet another identity—as a psychoanalytic psychotherapist. Not that I am entirely comfortable with that: the theories are so prolific, ever growing, unfalsifiable. You can't conduct an experiment to *disprove* any of them. So how can you choose which to believe, which to scorn? Hence I still find the Freudian approach so peculiar, quaint, scientifically questionable (or are these doubts just another form that my ever-mutating resistance is taking?).

Perhaps as a result, I still can't imagine becoming a psychoanalyst. I keep trying to envision that life—having an office on the Upper East Side, entering my own four-to-five-times-per-week, several-year training analysis. A life of anonymity, a stifling monasticism, it seems, bound to the Freudian (or *neo*-Freudian) mast for a lifetime. But so many of my residency classmates aspire to it. Annabelle and Georgie have already been accepted for training at the New York Psychoanalytic Institute; Zooey and a couple of others are starting night courses at the Columbia Psychoanalytic Institute uptown near the George Washington Bridge.

But still, I see undeniable signs of progress. Ms. J has changed as a result of our continued dialogue. She is better, and so perhaps am I. So despite my general reluctance, my constitutional skepticism and uncertainty about this whole approach, I'm starting to regularly look for transference issues in my work with patients and to try to identify my own countertransference feelings. I'm more comfortable with "the work," even in the most tedious and aggravating moments. And I'm less tortured by my own personal conflicts, less continually in crisis.

And whenever I feel bored or angry in the room with a patient, or particularly when I feel an urge to flirt, I hear Dr. Banks's voice: "Doctor, do you think there may be some*thing* you are avoiding?"

CHAPTER 2

Tigers in the Night

A Therapist's Own Therapy, 1981–1988

1

The front door is wide open; only the heavy wood-framed screen separates us from what stands outside. There, glistening and pacing, is an enormous monster. Huge muscles ripple under its striped coat. My little brother grins and—despite my vehement protests—opens the screen door.

I woke up in the call room at Payne Whitney, in a sweat, a nurse shaking my shoulder, summoning me yet again to some catastrophe I could do nothing to reverse. A patient had tried to hang himself. I ran down the hall after her to a room where two of the aides were leaning over a struggling man. He was gasping, fighting out of their grips, a broken shoelace still around his neck. I stood there, useless, as they wrapped him in a posey, a white straitjacket with leather straps that secured his arms behind his back, then ran him face-down along the hall into a seclusion room. This was followed by a shot of Haldol, which the nurse had drawn up as I wrote in the order book. After which I could do little more than watch him wailing and thrashing through the greasy wire-laced window. Then back to my call room, struggling to sleep.

Until this nightmare again awakened me, seemingly minutes later.

"Any ideas about the tiger?" says the man in the Mr. Rogers cardigan, sitting across from me with his legs crossed.

Silence. What am I supposed to say?

Before I knew the word psychotherapy, there was this dream—me at home with my elfin brother Jonathan, both of us little kids. I'd lie in bed, trembling, all alone in my attic room, looking at shadows of branches on the ceiling, afraid even to run for help.

It came back every so often, like strep throat or thunder. At first, I'd feel utter terror (*God! I'm about to be devoured!*), but later, when I was fifteen or sixteen, it didn't bother me so much; I almost began to look forward to its return, as strange as that might sound.

But not last night: last night the door was opening to Hell. Last night, every nerve thrumming with vibrations from the FDR Drive below, I lay paralyzed in that clammy hospital bed as the tiger leapt.

So here I am: age twenty-six, starting residency training, sitting in a high-ceilinged Central Park West office, facing a stranger in a red long-sleeved sweater, again fending off the tiger. It is September 1981. Now, just like when I was five or six years old, when I'd watch Jonny opening the door—sure of disaster.

From the outside, I'm doing well enough, I guess. Yet I'm totally utterly irretrievably miserable. It isn't just sleep deprivation from being on call; it's not just from being immersed in remorseless tides of illness and mortality. That's part of it, sure. But I make it to morning rounds, I excel at blood drawing and case presenting and all the other scut-work expected of an intern in an urban medical center (despite occasionally dozing off in teaching conference). I'm not killing people left and right. It's something different: *I have vanished.*

I have lost myself and have no clue where I've gone.

The tiger for the moment has assumed the identity of my residency advisor, whom I will call Dr. Firth. Perhaps a decade my senior, assistant to the director of residency training, on the surface a most un-tiger-like man, cultivated, politic, affable, interested in theater and contemporary art. Dr. Firth was one of the main reasons I chose this program. Surely a hospital employing such a paragon of culture would be an ideal place to become a psychiatrist.

Internship began in July 1980. I got through that year's medical rotations OK, then started psychiatry rotations the following July. And something happened to me. I got distracted, spaced out, focused on the wrong things, like the poetry of my patients' speech rather than what their problems were. I wrote stories in my head rather than listening. I forgot important tasks; I wasn't doing the assigned readings, or if I did, they seemed like meaningless jargon, words and paragraphs that could be replaced equally well with any other gobbledygook.

Within months, Dr. Firth has been utterly transformed. Still superficially amiable, asking at first with interest, then concern, and then with irritation that warns of rage, he wonders what is wrong with me. Why aren't you *there* with your patients? Your supervisors are complaining, and the complaints have a common theme.

Just what is it?

And so it is, like thousands of other psychiatry residents before and after me, that I backed myself into psychotherapy.

At this point I must make a disclaimer, or perhaps a clarification. What business have I, a psychiatrist, to write about my own psychotherapy? Rather than keeping secrets, why am I telling tales? According to the philosophy by which I am being trained in the early 1980s, a true believer (or as my supervisor Dr. Banks would put it, an *ethical practitioner*) never reveals his dreams to the world. Instead, you must pull a blackout shade over the secrets of your life. Things may have changed today, but back then, it appeared to be an ironclad rule: *Live one's life as a blank screen.*

At Payne Whitney in those days and in the 1980s analytic world in general, the therapist's anonymity in *and out* of the psychoanalytic hour is believed to be essential, to enable the patient to properly project his dreams and fantasies. The word is "abstinence." To know the analyst as a person is like sitting in a theater with two movies playing at once.

In practice, as a junior resident treating a few psychiatric inpatients, maintaining abstinence is not overly difficult: one merely turns each question, every wondering, back upon the patient: "*You ask if I'm married—what are your thoughts about . . . whether or not I'm married?*" Similarly, for *Do you have children? Are you gay or straight? Have you ever been depressed?*

Today, in an era of ever-present social media, the internet, and Google searches and given our more sophisticated understanding of transference—the interactions between patient and therapist are now seen as a cascade of unending interpersonal revelations—such extreme anonymity generally is felt to be unnecessary if not impossible, but back then the rules were ironclad. The system required a lifelong cloaking of one's self. To me, just learning the rules, this was stifling, entirely suffocating. I wanted to be a writer as well as a doctor, and how could I write about my own life and my own feelings if I needed to remain anonymous?

So, what do I have to say about that dream and how it connects to the catastrophe of my life? Barely noticing who sits opposite me (a stocky, middle-aged,

West Side version of Richard Dreyfus, had Richard Dreyfus attended Columbia Psychoanalytic Institute), I ramble on.

"Maybe . . . maybe the man's violence, the man who tried to hang himself, it got to me. So, the dream came back."

"Came back?"

"It's not new, I mean, I've had the dream before."

The stranger looks at me. "When did you start having it?"

"When I was a kid. . . . Here's the thing, the tiger is beautiful, but it's going to kill us."

"Can you say more about that?"

"Maybe that's the tiger, this mess I'm in." Why did I even go to medical school? Why did I choose psychiatry? Why had I chosen this psychoanalytically oriented program in which it appears essential to plumb your own depths endlessly? It all seems crazy. "I was so sure about my choices, but now I don't know. I may have made a terrible mistake—a whole series of mistakes—and I can't go back."

But then, just as I start to catch my breath, Richard Dreyfus says, "Sorry, our time is up."

That phrase—is that *actually* being spoken aloud? I laugh, awakened from my many nightmares.

"Would you like to set up another appointment?"

What can I say?

So it is that I begin my work with Dr. Veltrin. Through the fall and winter of 1981, I rush out of the hospital every Friday after morning rounds, and then every Tuesday and Friday, sometimes signing out, other times sneaking away and praying that my beeper won't sound.

I tell almost no one that I am going: certainly not my parents, not my friends or my fellow residents (though many are in full-fledged psychoanalysis themselves). I only reveal it to my girlfriend Lisa after several weeks. There are a hundred reasons for that, I suppose. Embarrassment for one, feeling that such a need indicates weakness, admission that I'm not in control of my life. Being terrified to be dependent on anyone else. And perhaps above all, the very unscientificness (is that a word?) of it all—I could only imagine what Dad would say about it, the whole method of talking about whatever comes into your head. Not to mention what he thinks about my choice of this field at all.

I do notice that Dr. Veltrin is generally benign and mild, except for the occasional wry turn of phrase, an impatient clearing of his throat if I head in a fruitless direction, or a shifting in his chair if one of my frequent half-dream-like silences extends too long. To be honest, I can barely ever recall any of those mild, calming things he says. They don't survive the wakening. In the sort of half-awake state of the therapy session, they make sense, but then I jolt back into real life barely remembering what I've figured out in the hour before.

2

The ashtray in my office, filled with stale odors, dead cigarette butts, somehow I've been dropped into it, I'm standing inside a giant black plastic ashtray with nauseating smells, surrounded by gigantic soggy smashed cigarettes. Then I'm in water, in the river, the East River that rushes in front of Payne Whitney, bouncing amid the standing waves of Hell Gate.

"So, I'm drowning, obviously. And probably getting lung cancer from all the smoking. But so what?"

"You don't want to talk about the dream?"

"No." A long, almost endless silence. I am in a lousy mood today.

"What? Can you tell me what you're thinking?" he asks eventually.

"I . . . I don't know. I guess I want to know whether psychoanalysis works."

I tell Dr. Veltrin about yesterday's conversation with my supervisor, a dour New York Psychoanalytic Institute analyst. We were in my office on the locked inpatient unit, where I was working at this point, my second year of residency. A high room whose thick mullioned windows looked out over the glare of the FDR Drive and the roiling December waters of the East River. I asked him straight out: "*Does psychoanalysis work?*"

He was appalled. "What one might better ask," he had responded caustically, "is, '*How* does psychoanalysis work?'"

"And what was your feeling about that answer?" Dr. Veltrin asks now.

"I thought it was stupid. And obnoxious. If it doesn't work, who cares *how* it works? But . . . I didn't realize how full of shit that was until after he left."

"Of course."

I look at Dr. Veltrin. Every so often, he says something annoying. What he means is, *of course* I didn't think of a good answer until after my supervisor left. I didn't even realize how I *felt* until he was gone. Often what Veltrin says is annoying because it is true: these days I'm never able to give good answers until it's too late.

"But I meant just what I asked."

"*Do* you think psychoanalysis works?" asks Dr. Veltrin.

"*No.*"

Dr. Veltrin seems almost pleased by my answer, which annoys me even more. But that's the truth. I don't think it works.

That year, psychoanalysis (or psychoanalytic therapy, or whatever it is I am doing) definitely does not work, at least for me. Endlessly, Dr. Veltrin and I discuss my relationship with my father—a World War II veteran, a son of immigrants, who grew up during the Depression, a wonderful, complicated, and overwhelmingly infuriating man—yet nothing changes. When I meet Dad at home in Ohio or in New York, when he passes through on his way to or from a medical conference, I am no better able than ever before to interrupt his diatribes. On his way to a World Health Organization conference in Africa or China, he'll stop by my apartment on the Upper East Side, lugging his black suitcase and wooden case of lantern slides, with a few hours to kill before a taxi whisks him to JFK Airport, so we grab dinner at some Chinese restaurant on Second Avenue or breakfast at an East Sixties Greek diner. He's forceful and endlessly opinionated, with his slicked-back black hair with a few gray strands and a vehement way of talking, basically uninterruptable, always telling me what to think, what to do. One way or another, it all comes back to this: Why don't I start doing research, which is the only thing that matters? "PET scans," he says, like the guy in *The Graduate* who tells Benjamin Braddock about plastics. Every time he comes to our apartment, he pulls the same towel rack off the wall in our bathroom.

"Let me talk! How do you know what I think?" is always at the tip of my tongue, but I rarely speak up.

I do begin, however, to see myself with painful clarity. I become excruciatingly aware of my need to excel, to vanquish my classmates with my accomplishments, just as I do with my father, while at the same time (certain of my coming disappointment) making sure to thwart and frustrate anyone who expects anything from me. I'll pretend to be totally indifferent, while inside I'm starving.

I have been seeing Dr. V (as I begin to call him) twice per week since the fall of my second year of residency, yet nothing is changing. If anything, I'm more frustrated, more infuriated, and more trapped than ever.

Sure, I can understand how in moving at the age of ten from the second-floor bedroom I shared with my younger brother Jonny to my own room in the attic I became isolated from my five sisters and brothers. I can understand Dr. V's theory that, as one baby was born after another, how my mother (eventually with six children aged ten or less)—*maybe the tiger at the door was my mother bringing yet another baby home?*—became overwhelmed and had so little to give to me and how I developed, in compensation, a magical life in my attic room, a life of words and dreams, and an intense (sometimes warm, other times rageful) relationship with Dad. I spent a childhood hiding under the eaves of the house. It makes sense; it gives me a different perspective on things. I feel sorry for my mother, with six little kids at home; I feel even more sorry for myself as a forgotten kid in the attic, unable to get her attention or so many other things I needed.

But even with these understandings, it seems impossible to encompass the massive weight of history in my family. Things that are beyond any therapeutic conversation fill our lives. The way that sixteen thousand of our relatives in Brest-Litovsk were machine-gunned to death in the Bronna Gora forest over two days in 1942. The way millions died less than a decade before I was born. As a World War II veteran wounded in combat, as a liberator of Bergen-Belsen, Dad's single-phrase explanation for having six children was "one for each million."

How can Dr. Veltrin help with that? I wondered that then, and I wonder it now. I had and have no shortage of doubts. But however impossible, it didn't stop me from trying.

"In a way," Dr. Veltrin says, "All six of you grew up needing to make up for the ones who died. That seems like an impossible burden." *I understand, but so what?* If anything, my troubles are worsening.

In February 1982, the middle of my second year, Dr. Firth calls me in. Again, my supervisors' ratings are consistent: they see me as distracted and preoccupied, and they wonder if I'm committed to life as a psychiatrist. I seem angry, too: I yelled too loudly at a nurse who didn't fill an order on the inpatient unit, so loudly that she reported me to the unit chief. I'm always making sarcastic comments in the psychoanalytic theory and psychopathology classes. (This is the least of it; I'm also finding myself pushing people on the subway if they won't move fast enough when I'm getting on or off and yelling at one of my fellow residents who won't pay back a call night she owes me after I covered one of her shifts.) Firth just sees me at work. But if you look at my entire life, I'm totally losing it.

"Have you thought of going into therapy?" Dr. Firth asks. "That might help you sort things out."

I stare at him. Are things really that bad? "Oh, yes, I've thought about it," I say.

Truly, the training of a psychiatrist is unlike that of any other doctor. After all, the surgeon's apprentice does not go under the knife; the pathologist does not offer up parts of organs to the microtome; gynecologists do not submit themselves to pelvic examinations; anesthesiologists do not go under the gas. Only the psychiatrist, learning psychotherapy—especially in the twenty- or thirty-year reign of Freudianism—is made subject to his own craft: He learns half upon the couch.

In the 1940s and 1950s, the post–World War II heyday in America, any psychiatrist worth his salt dreamed of qualifying to become a psychoanalyst. One had to bear the rigors of applying to an institute, of having one's personality minutely scrutinized for possible defects making you "unanalyzable"—for instance, having a character disorder or being gay. Homosexuality was a pathological disorder, a reason to be expelled from psychiatric training and practice. Those few favored to gain acceptance next began a grueling apprenticeship. Years of classes in psychoanalytic theory, nearly cost-free treatment you provided to several-year-long psychoanalytic cases from start to finish, and—perhaps most importantly—your own "training analysis," which often involves starting psychoanalysis over with an institute-approved psychoanalyst.

By the early to mid-1980s, when I am torturing the training system, the psychoanalytic movement had fallen on hard times (the august New York Psychoanalytic Institute is close to bankruptcy), and few psychiatrists outside of New York and Boston still want to become psychoanalysts. Nevertheless, it seems that the Payne Whitney residency almost invariably precipitates some kind of personal crisis, such as I experience with Dr. Firth, and a push, gentle or otherwise, into a "personal psychotherapy."

The ideal, of course, is if you embrace full-blown psychoanalysis. Most of my fellow Payne Whitney residents, I gradually realize, have done so already, seeing their shrinks several times a week. You lie on a couch for fifty minutes at a time, three or four or five times a week; you say whatever comes into your mind; your analyst sits at the side of the couch, behind your head, so you don't see him or her. Mostly you are enveloped by silence, as you talk, your analyst saying a few words to your many. "Can you say more?" "How do you feel about that?" "What are your thoughts?" Thoughts of all sorts emerge in such circumstances, uncensored wishes and fears and fantasies, deeply buried memories rise up, as do intense feelings of all kinds. You work day after day, hoping to gain a deeper understanding of your unconscious mental life, and to "work through" any unconscious difficulties that would interfere with your ability to work as a psychoanalyst and that also interfere with your ability to live fully in the world as a person, an adult, a parent, a child, a sibling, a worker, a lover. "Some good work" can be done with three-days-per-week analysis, they say, but five times a week is optimal.

Most of my classmates plan to continue on to psychoanalytic training, and the chief resident almost always seems to be a budding analyst. *Here*, at Payne Whitney, that is, being in full-fledged analysis during residency is a mark of the elect. Not only are you suitably cerebral and intellectual, but merely bearing the cost of full psychoanalysis proves you've got the right stuff: the monthly fees for many-times-weekly sessions—even if at a discounted rate for analysts-in-training—are greater than our take-home pay.

All through residency training we joke about the foibles of our professors and supervisors as proof to the adage that psychoanalysis doesn't cure your neuroses, it just frees you to brag about them.

Analysis—is it for me? I try to imagine four or five days a week with Dr. Veltrin. Money is only one part of it. One weird aspect of Payne Whitney is the privilege conferred by psychoanalysis—if you're known to be in analysis (and somehow the word gets out), you become an automatic insider. I contend that I want to be in treatment for its own sake, not for where it might get me.

And there's something else too: tameness. Classmates who are in full-blown psychoanalysis rarely raise a fuss about hospital conditions or anything else. Infinitely self-reflective, they're embarrassingly eager to please. Practically brainwashed, it feels.

But . . . are these observations valid? Or are they just my ever-mutating "resistance"?

Spring of my second year of residency (our training years start in July, so my second year ends in June 1982), utterly miserable, I make an appointment with Dr. Firth. I want out, I tell him. I need to take a year off, to catch up on my sleep, to get my head together, to think, to write. I'm not enjoying my work. Being a psychiatrist is too damn hard. Doing psychotherapy seems impossibly difficult and fruitless. As far as I can tell, therapy doesn't work. Not for me, not for my patients. So why continue?

Actually I don't say all that to Dr. Firth, but I'm sure I convey as much. Firth listens patiently enough, and then responds that it is OK with him if I want to take time off beginning in July, the start of the next academic year; he'll OK it with the director of residency training. He pages me the next day to confirm that time off is approved, but I must let him know definitely one way or the other within a week, because other candidates might want my spot.

Two weeks later I am mightily surprised to get an irritable call from Dr. Firth: Have I decided?

"Decided what?"

Pure rage bursts through the phone line. How can I not remember my promise to get back to him? Don't I know I am keeping someone waiting? How inconsiderate!

Totally humiliated, I stammer out some excuse.

"Well, what's your answer?" Firth demands.

"Oh," I respond in the apologetic tone I seem to be forced to take so often those days, given my inexplicable amnesia. "I've decided to stay."

Odd, in view of my nihilistic views of the efficacy of psychotherapy, that I don't quit seeing Dr. V at that point. Odd too, in view of my rage toward the training program, toward Dr. Firth and his colleagues, a rage that has spread like blood through water to wash through every moment at the hospital, that it does not extend (except for brief flashes) toward Dr. Veltrin. I know (in terms of transference) that I am "acting out" my anger at work (to my own detriment, usually), that I *have become* the tiger, that in letting Dr. Firth wait in ignorance of what I might decide, I'm likely playing out some drama of my early childhood, of repeatedly being kept waiting as one of many children in a household where it often felt like there wasn't enough to go around. If therapy was going right, shouldn't I be having more transference feelings toward Dr. Veltrin himself? Shouldn't my rage, at least sometimes, be directed toward *him*, not toward Firth? Veltrin has assured me he can take it; that's what he's being paid for. While Firth, trying to teach, advise, administer, has barely a clue.

Yet perversely, perhaps *because* I know these feelings might develop toward Dr. Veltrin, they just haven't. By this point I feel comfortable with him. He seems like a really nice guy. His observations and comments are making a lot more sense. It's just that, however true they might be, they don't seem to make any difference. (In retrospect, though, "nice" should be a clue: Dad always speaks scornfully of the "nice fellows" he knows, practitioners around the medical center who get on so well with patients but who really don't know a damn thing. Back home in Ohio, "nice" is an insult.)

I do find myself thinking about how one of my supervisors during my medical school days said psychotherapy is a Zen process, drops of water falling on a stone, which will eventually wear a new path. Perhaps, I muse, water isn't enough for me. Sulfuric acid, maybe. That fall I begin to see Dr. Veltrin three times a week.

3

Incessant dreams set on boardwalks near the ocean, among wind-sculpted low trees and dunes of the sunken forest. Wild chase scenes, all-terrain vehicles zooming over sandy hummocks, crackles of gunfire, diving off into the undergrowth, lying motionless for hours. Once there are even hideous monsters, perhaps tigers, springing from the brush.

What do we talk about all those days?

Sometimes it's all about my wild years, when I dropped out of college and hitchhiked through California, ending up in San Francisco, working as a bicycle messenger for the Speedy Delivery Service. I was living in the Castro District, rooming with a hippie couple and a pregnant secretary and escaping everything back East, dodging cabs and buses, delivering packages up and down the steep inclines of Russian Hill and the North End, enduring the Speedy dispatcher hitting on me, finding myself stoned in dank booming art lofts or wandering through redwood groves, somehow getting invited to parties at communes located at abandoned silver mines, wild days during which I was convinced I'd never go back East, never finish college, just stay a dropout for life.

Sometimes it's about why I came back. Somehow, I wanted to be a doctor. And I felt that if I never came back I'd miss something essential. That wanting surprised me; it snuck out of nowhere, the wanting to heal. Which led to this, the decision in the last year of medical school to become a psychiatrist, where I had somehow made an irrevocable commitment to a constrained and strange existence—a novitiate in this weird analytic Manhattan priesthood—my own wildness somewhere still within, with nowhere to go. Except toward trouble, day in and day out.

Hence (at least at work), one crisis after another. Confrontations, mixed evaluations, fights. Supervisors read their evaluation notes to me: *Needs more psychotherapy supervision: seems preoccupied, distracted. Conflict with charge nurse. Nonresponse to beeper with walk-in patient.* Even my writing gets me into trouble: I have some articles published in magazines, and an administrator stops me in the hospital lobby one day. "Who gave you permission to write?" she asks accusingly. "Does the chairman know about this?" Clearly, I am a total mess. How can I possibly be a good therapist unless I get this all sorted out? To be a competent psychoanalyst, your internal life cannot be chaos. Or so the theory goes. And how can I disagree?

I arrive at Dr. Veltrin's office, get buzzed in, and sit in the waiting room, perhaps sipping deli coffee from my blue-and-white paper cup or flipping through last year's *New Yorkers*, until the double doors open and the previous patient emerges—sometimes a fellow Payne Whitney resident—and sheepishly dons his coat and departs. Then Dr. Veltrin appears at the door ("C'mon in"); I enter and settle in the still-warm chair.

After completing my recitation of the past week's disasters, I sink into a separate place: not precisely the world of imagination and not precisely the world of the subconscious (after all, I'm fully alert) but some sunken world, some unique vantage point from which all the chaos of day-to-day life suddenly resolves into order, into patterns and shapes and forms, not necessarily the ones I want it to have but ones that are unmistakably true.

An entire world, yet one whose textures and colors often fade right after I rush out the door onto Central Park West to grab a taxi back to the hospital.

Certain things have become clear. Many things make me squirm in that chair to which I don't give a moment's thought the rest of the week. Some moments the whole sky lights up, like in the middle of a thunderstorm on a lake in Maine, when everything suddenly glows brighter than day and you wait for an enormous boom.

Competition, for instance. It becomes clear my rabid competitiveness with my fellow Payne Whitney residents reflects not only my relations with my brothers and sisters but also my father's horrid experiences dealing with "Jewish quotas" for medical training in the 1930s and with Nazi-sympathizing Bundist medical school professors, one of whom fled to Mexico after the start of World War II. Control, for another. How in fights with my girlfriend Lisa I tend to withdraw, to retreat to the attic of my mind, and find myself actually *unable* to talk, mortifyingly silent, emotionally vacated. Rather than letting loose and saying what I really think, my mind entirely empties of thoughts, which embarrasses and almost horrifies me. Yet when I am able to force myself to talk even after interminable silences, things come out that make some kind of sense, however painful. And progress ensues. Certainly that's true with Lisa, who moves in with me after finishing graduate school: our fights have noticeably improved. After such silences, I am able to talk.

In summer camp as a kid, pushed into a boxing ring, my punches scared the other boys, my left hook surprising them with murderous intensity. In college, in organic chemistry, I nearly failed the first quarter, then at the semester final exam, went from a C− grade to the highest mark in the class. I was prone to sudden collapses, then a reemergence somehow even stronger, often on top. There was a French phrase that kept coming to mind: *reculé pour mieux sauter* . . . during my endless silences in the room with Dr. V, all of these things seem somehow connected, emerging with strange intensity.

As my residency continues, I start hearing accounts of apparently successful therapies among my fellow Payne Whitneyites. It's amazing how prosaic such observations can be. "I realized Dr. X was just another person," reports a young faculty member who just finished analytic training, describing six years of four-times-per-week psychoanalysis. Zooey, a fellow Payne Whitney resident, confides, "I learned a lot about anger." My friend Annabelle tells me, "I realized my mother was too dependent." It is as though Sir Edmund Hillary descends from Everest to report, "Awfully bloody high." Or Livingstone, emerging from deepest Africa, croaks out only, "Wretched bugs."

Is it mere amnesia? I wonder. Does it prove the essential triviality of psychotherapy? Or does it perhaps reflect something else: the difficulty in translating that night-world back to the language of quotidian life? Try explaining how you spend your workday to the tigers in your dreams! But, for such therapy to work, though, don't you have to build some conduits between the worlds of life and dreams? If so, I am resisting full strength.

As a psychoanalytically oriented therapist (but certainly no psychoanalyst—that requires years of additional training) I still feel like a rank beginner. I still get lost in my patients' words, their stories, but now I struggle to connect what they are saying to their emotions, to their conflicts, to their feelings toward me. At times I am able to do a bit of this, with my first long-term exploratory therapy patient, Ms. J, and with a sixtyish grocery store supervisor with bipolar disorder, who feels comfortable telling me how his mood swings affect his ability to work with customers and how he struggles to keep his job. And with a scared young woman with hebephrenic schizophrenia who is brought in by her elderly father, both of whom thank me profusely: they say they trust me, they like working with me, they can tell me things they weren't able to say to their previous doctor.

In continual crisis, I manage to struggle through my third year of residency, working on a consultation psychiatry service with burn patients through the end of 1982. Then in January 1983 I start in the ambulatory services, seeing reduced-fee patients in the basement of Payne Whitney.

That July, I begin my fourth year as an assistant chief of an inpatient service, and then for some reason, things begin to ease up. I say "for some reason," but, really, I do have an idea why it is. It is because I have begun feeling . . . *better.* I'm starting to enjoy my work. I even start wondering again about life as a psychoanalyst. My classmates start applying to psychoanalytic institutes. I see them jumping chummily into cabs with supervisors and professors, on their way to evening events or called into special meetings at Payne Whitney for "incoming analytic candidates"—I'm both relieved and oddly jealous. I understand why they want that life, but fundamentally, I gradually realize, I don't. Not only "don't," but in a way, I "can't." Temperamentally I couldn't live that life: that I realize with increasing clarity. Yet I still feel envy—they belong, I don't.

After much discussion, Lisa and I decide we want to stay in New York. Having decided not to become an analyst, it is clear that I will leave Payne Whitney: but where will I go? That fall I apply for a fellowship. I finish a big research project and am surprised to get the "Best Resident Research Project" award from my department. Then finally, in June 1984, I graduate. Lisa and I buy a rooftop Manhattan co-op, a wreck, former maids' rooms conglomerated into a two-bedroom apartment, surrounded by a huge terrace (this being the good old days when this is still possible for graduating residents). I ease into my fellowship program at the New York Psychiatric Institute, way uptown near the George Washington Bridge, spending half my time at Manhattan Medical Center in lower Manhattan. At the end of that year, I begin looking for a permanent job. I finish two books and am amazed when both are accepted for publication.

Despite all these changes, despite various signs of conventional success, the tiger still threatens every day. Some things are undeniably better. I have it out with my father on a couple of occasions, and things begin to thaw between us, sometimes even becoming friendly. He sends stacks of reprints made by his secretary, Mrs. Husselman, along with photocopied letters from his MD colleagues who have read my publications—and his proud responses back to them. He comes back from a second trip to Africa, and we go for a walk up Fifth Avenue, ending up at the Temple of Dendur at the Metropolitan Museum, then buy hotdogs and Cokes from the vendors out front and sit on the steps and chat. I'm doing better, feeling less miserable, less suffocated than during my residency, yet I still feel as though I fight off one crisis successfully only to encounter another coming around the bend—like James Bond emerging from a pool of sharks only to find himself in a collapsing chamber, from which he escapes only to discover himself surrounded by Spectre agents with guns drawn. I'm still always on edge: there's always another issue coming up.

"Y'know, you hardly ever really look at me," Dr. Veltrin says one afternoon.

A long silence as I consider what he has said. We have talked for month after month, we have worked together to create an intricate lattice of explanations, of feelings, of family connections. I have learned things about him, both directly and through the professional grapevine. A random collection of odd information, but more of what I know is from being in a room with him for hundreds of hours. His daughter was in the office one morning, a slim young woman with two large dogs on leashes, for instance. He lives in a fifteen-room apartment, says the Payne Whitney grapevine (which incited enormous feelings of envy in me). Sometimes I run into him when jogging around the Central Park Reservoir: he seems to prefer running clockwise, the wrong way, whereas I run counterclockwise, so we pass each other twice each lap. More recently, there's new gossip: he's having marital problems, someone says; he's leaving his wife.

And it's true that I don't often look at him. I usually focus on the fireplace behind him. Now I look at him. A strange set of feelings comes over me, and it takes a while to put them together. I trust him. Is it as simple as that? After, what is it, more than five years? I can say whatever I want here; it is safe. I find myself in tears.

One night I dream that *I am on a sailboat on a highway, a kind of high-tech boat on wheels. An exhilarating wind comes up, and I zoom along the highway, passing cars left and right, doing ninety-five.*

4

It's hard to document healing, especially when there are no fractured bones or visible scars to show. To point to improvements in one's life can sound like a rationalization. Don't we always tend to justify what we have done, however boneheaded, if only because we have done it? Besides, who is a good judge of cause and effect in his own life?

Do these changes result from my psychoanalytic psychotherapy with Dr. Veltrin, especially the recent conversations that finally turned around my relationship with my mother? *You so rarely talk about her . . . why do you think that is?* Or merely from growing older, becoming better adjusted to life as a doctor, as a husband, as a man? From my life as a new father? For on a cold February day in 1987 our daughter is born, and our lives change forever.

Whatever the cause, I am definitely, finally, changing. Unsolicitedly, people tell me I am more open, more emotionally available. That I seem more "there." Patients open up more easily, tell me things they've never told anyone before. After a year of almost no patients, my private practice finally begins to grow. At the hospital, students want to work on my research projects. Job evaluations turn positive, and for the first time in memory I get on reasonably well with my superiors.

The only place I am floundering, oddly enough, is in my writing. The sharp edges of my previous responses suddenly seem put-on, not wholly credible. In reading my old work I cringe: *I really said that? What was I thinking?* And in trying to do new work, I find myself pausing: now what?

"You're always reporting what's happening from the outside," Dr. Veltrin had commented a year or more earlier. "You're very attuned to the external world. Don't you think it would be better if you had more access to what's going on inside? If you could use more of that when you wrote?"

"No," I answer, no doubt a bit too quickly. "You have to have edges to have a style." Look at Hemingway, I say.

Now, in a literary slump (and at the same time strangely indifferent to so many issues that had preoccupied me since I was a teenager), I begin to reconsider: Maybe too much energy is still going into trying to keep the screen door shut. That's what we talk about today. My impasse.

"Maybe," says Dr. V. There's a pause. "You know, you don't really seem like a Hemingway. You seem more like a Cheever. Someone who writes more about emotions, of people living in different subcultures, different worlds."

"Maybe," I say. "Maybe so. And that's much harder."

"And maybe that's good?" he says. "But you seem angry."

"Well," I say, "Maybe . . . it's because I don't feel it's been 100 percent successful here. Seventy or 80 percent maybe, not bad as far as these things go, but . . ." I had hoped for more, some sort of transformation, not just this.

A pause. The garbage trucks have arrived early this Friday morning and are making their usual racket on Central Park West outside Dr. Veltrin's windows. I'm feeling at once both great sadness and relief. I'm not sure where it's coming from. Now that I'm thinking of ending therapy, a lot of strange feelings are emerging. It has been so many years. Things are not resolved. Like they say about a poem, never finished, only abandoned. A strange burst of anger a couple of weeks ago, criticisms leveled against Dr. Veltrin. And it's not in a vacuum, either.

I don't recall what else we talk about that day, just that all too soon it's time to go out to Central Park West to hail a cab, impossible at this hour of morning. Finally, a beat-up lemon-hued Checker, the high, rumbling old kind, pulls to the side. I open the door, confronting savaged black upholstery. The jump seat, wrenched from its moorings, wobbles on the floor.

"Hey, are you a cab, or what?"

The driver turns, scowling. "Buddy, this is a South Bronx taxi. Hop in."

The trees of Central Park flash by, and we veer down Broadway toward my hospital, where I am now a clinic doctor. It is the spring of 1988, the midst of the AIDS epidemic, with over one hundred thousand deaths in the United States alone. We are overwhelmed at work but oddly optimistic. We are doing great work.

I recall a dream from last night: *On the third floor of my parents' house, in my old attic bedroom, I discover a door beneath the eaves of the roof, where in waking life I know there's only a storage compartment. There I discover a large room that leads to another room and yet another. A whole wing of the house whose existence I had never suspected. I come out on the roof and discover that it is night, and the sky is bursting with stars.*

I grin. How utterly obvious. New rooms opening up. New opportunities. *The sky!* Barely worth recounting next week. But damn, I realize as we roar downtown, I *am* mad at Dr. Veltrin. Really annoyed. Don't know what that's all about—maybe I want him to be more perfect. A perfect psychoanalyst. A perfect parent. Of course, the issue must be how good of a parent *I'm* going to be. And how good of a psychodynamic psychotherapist I'll be.

In front of my new hospital, far downtown, blessedly far from Payne Whitney, a place far more hospitable than I could have ever imagined, the South Bronx taxi disgorges me. I lean over to pay the fare and see a huge machete in a leather sheath on the front passenger seat. My heart pounds.

He rumbles off, stopping abruptly before another citizen, an elegant woman. *"Hey buddy, are you a taxi?"* I hear. Should I warn her? Too late—she slides in, they take off downtown.

I half run, half walk toward the clinic, late for morning conference. They're waiting for me.

CHAPTER 3

The Enchanted Garden

Psychoanalysis in the Psychiatry Marketplace, 1985

On one level, you might say that annual conventions are all alike. You have the plastic badges, the open bars, the tipsy bonhomie of innumerable salesmen. And the slapping of backs, the unexpected collisions of colleagues who haven't seen one another since the last shindig, and the crowds that wander dazedly through vast echoing halls from one extravaganza to the next, filing onto Greyhounds, and placidly dozing in dank amphitheaters. As Alexis de Tocqueville noted two centuries ago, we Americans are in love with our associations. And associations have no greater moment than their annual conventions, whose grandest purpose is to amplify the superiority their members feel in comparison to the democratic swarm.

But this one, the American Psychiatric Association (APA) Annual Meeting? Clearly I'm wandering into a circus.

It is May 20, 1985. I graduated from residency training less than a year ago and have now come to Dallas to present my research study on command hallucinations, which will be in a poster session tomorrow. This is a weird thing in itself: how I ever decided to do research, despite Dad's pressure to do it. My opposition to Dad's constant pushing somehow became beside the point when I found out there was a database on several thousand patients admitted to Payne Whitney and realized I could do a study of command hallucinations, of voices that tell psychotic patients to act in a particular way, often to hurt themselves or others. Something came over me, something emerged from me: a passion for studying things, for doing research, regardless of what Dad might want for me, from me.

In a matter of weeks, in my last year as a resident at Payne Whitney, I managed to assemble a group of advisors from among the faculty and wrote a computer program in Fortran and BASIC to process the data—a stack of IBM punch cards that you fed into a Buick-sized data processing machine. I got my classmate McNeil to help me code the data, and then once I had results, I wrote it up and submitted it for presentation at this APA Annual Meeting.

To me it seems that we need Hunter S. Thompson here, or Ken Kesey, or Tom Wolfe at least. This is not just another convention; this is somehow the mash-up battle for the future of the mind.

Which might explain my odd excitement and apprehension as I sit on a couch in the vast lobby of Dallas's Loew's Anatole Hotel, paging through a telephone-book-thick schedule book, circling promising lectures, seminars, videos, and parties of the 138th APA Annual Meeting and watching my fellow psychiatrists convene—proof that I may possibly be able to join a larger world than the psychoanalytic hothouse where I trained at Payne Whitney, the place from which I have just barely escaped.

"Hey there!" It's McNeil. McNeil from New York, one of my poster coauthors, my buddy. We endured four years of residency training together, finishing just last June. Since then, I've been doing a public psychiatry fellowship at the New York State Psychiatric Institute, with a clinical placement at Manhattan Medical Center, a rough-and-tumble downtown hospital; he's taken the usual route for Payne Whitney grads, opening a private practice and starting psychoanalytic training. Although we're good friends and live only a mile apart in Manhattan, our paths haven't crossed since graduation.

McNeil sinks on the Naugahyde beside me. "What are you going to do today?" he says.

"Don't know," I say.

"There's a lot of good analytic stuff," he says, flipping open the paperbound conference proceedings: *Theoretical issues in psychotherapeutic practice. How to set up a therapy contract with a borderline. How psychoanalysis has changed over the past half century. Screen memories and narrative truth.* "At twelve-thirty they have Psychoanalysis and Psychoanalytically Oriented Psychotherapy. My analytic institute supervisor is presenting there."

"Boring!" I say. "What about the workshop on How to Survive a Television Interview?"

He laughs. "Hey, have you been down to the convention center yet? Lots of great freebies there." He empties a shopping bag on my lap. Notepads embossed with the names of antipsychotic medications, squeezable rubber brains with drug company logos, flyers from antipsychiatry activists.

"Nice clipboard!"

"Got the last one. Oh, and don't forget to pick up your tickets, either." He fans a thick wad of tickets to drug-company-sponsored events: Smothers Brothers in concert, a reception at the Dallas Museum of Art, the Dallas Chamber Orchestra, the Mesquite Rodeo, an over-the-top dessert party at South Fork Ranch. "Y'know—that's where they film *Dallas*."

"Wow!" It's beyond strange, emerging from the insular world of the Payne Whitney Clinic, my psychoanalytic reverie-turned-nightmare, to the reality of this brave new world of swag and circuses.

"But they're all out of those." He sighs. "God! So much to do today! Still haven't written my case for the psychoanalytic symposium tomorrow."

"Really? I still have to do the power analysis for my poster."

"You better get cracking. Hey, you coming to the party?"

"Which one?"

"You know, the Payne Whitney blowout, where they spend ten thousand bucks on shrimp and crab and booze. Tonight, at some mansion downtown."

"I'll try to be there."

"Great!" He glances up. "Oh no." A short, plump woman crosses the hotel lobby, approaching us.

"She's a psychologist, was in my sex therapy course. Been following me all day. Gotta go, bye."

Pretending to read my program schedule, I watch him duck through a crowd of natty psychoanalysts, the woman ambling after him. Then I get into my rental car and follow the maze of freeways downtown.

Immediately I get lost. I find myself veering toward Fort Worth, and a few miles later, I loop-the-loop on a cloverleaf and hurtle through downtown Dallas. Squinting at the map, I try to divine which building houses the Texas Book Depository, where Oswald crouched when he murdered JFK, and which exit to take before I'm catapulted southward toward Houston.

I park amid expanses of concrete and glass. Men rush along the sidewalk, one holding a cardboard carton labeled "Uzi," with the photo of a submachine gun on its side, two others holding naked automatic weapons. "National Rifle Convention" reads the sign overhead. I feel a surge of panic.

Then I see the flapping blue-and-white banner: "Dallas Welcomes the American Psychiatric Association," beneath which doctorly types converge, sporting blue plastic briefcases and green-and-white nametags, identical to my own, and I can breathe again.

Ah, America!

After registering, I wander through the echoing convention hall. It's buzzing with psychiatric jargon, crowded with computerized billing systems, clunky electroshock machines. Video monitors blink scientific data, book publisher kiosks display everything from the handy new yellow-jacketed pocket-sized mini-edition of the *Diagnostic and Statistical Manual*, 3rd edition (*DSM-III*) to leather-bound first editions of Adler and Freud. But mostly it's one Big Pharma display after another, each gaudier than the last, their salesmen lying in wait, eager to chat you up.

And there is a strange excitement on the floor of this vast hall today: a new drug being used in Europe. Fluvoxamine. First of a new class of "selective serotonin reuptake inhibitors," or SSRIs. Whatever that means! Supposedly safer than the tricyclics and MAOI antidepressants we currently use, with hardly any side effects. Being as they're not yet FDA approved, no one can actually talk about these SSRIs, least of all the salesmen. So, it's all whispers, all buzz.

"Fluvoxamine should be here soon," they whisper, meaning available in the United States. "Officially, we can't tell you that, but you can read it in the journals."

First of class, maybe to turn our world upside down. Or is it all hype? Haven't we heard this before? Haven't many other drugs piqued our interest only to dash our expectations? I stumble down a booming hallway, reminded of the line from William Carlos Williams's *Autobiography*: how the modern hospital "is part of the fairground for the commercial racket carried on by the big pharmaceutical houses."

And here we are smack-dab at the center of the fairground for the mind and the brain. Maybe it is all hubbub—or maybe, amid the racket, could there actually be a glimpse of a whole new universe?

I take refuge in "Brain Imaging: Clinical Implications." The lecturer flips through computer-generated color brain maps, dazzling us with regional cerebral blood flow, PET scans, computerized electroencephalograms, with nuclear magnetic resonance images that pulse on the screen overhead. So this is what Dad is always going on about. How come he, as a cardiologist, knew about this before me, the psychiatrist?

Sitting there, gazing at brains displayed in a rainbow of colors, you can easily forget that this is *psychiatry*. Which leads me to wonder even more: What does all this have in common with the talking cure that McNeil and I have just spent four painful years learning? With object relations and projective identification at the center of our curriculum? With all the transference and countertransference stuff we struggled to master in thousands of hours of psychoanalytic therapy and supervision? The researchers standing before this darkened room might as well be radiological necromancers.

Exhilarated, profoundly confused, I wander back out to the main hall. Antipsychiatry activists have set up a table piled high with flyers; nobody's talking to them. I feel kind of sorry for them, but I'm not really in the mood for castigation, so I go back to the drug salesmen, to their fistfuls of pens and colorful notepads. *Asendin . . . Like Helium for the Mind.* McNeil was right, I discover. The free clipboards are gone. So are the tickets to South Fork Ranch.

Around five, on my way to the Payne Whitney party, I get lost again, in the blank, pedestrianless Dallas streets, but finally I find the skyscraper-dwarfed mansion. I park around back, grab a margarita from the porch bar, and wander inside to a miraculously reconstituted fragment of Payne Whitney Clinic.

Virtually all the powerhouse psychoanalysts are there, my former professors, sipping drinks, and so are my old friends and rivals from residency, mostly graduated to private practice and one or another of the psychoanalytic institutes. Their predecessors, Payne Whitneyites from past decades, older and stockier, many sporting gray-specked beards, have flocked here from all over the country.

"This house," proclaims Dr. Firth in grand mode, perhaps slightly drunk, "used to be a bordello, then it was a funeral home. And now it's about to be torn down for a bank skyscraper!"

It's weird how everyone, all these psychoanalysts, listens even to his blather with such incredible seriousness and calm intensity. It's supercharged, this vast crowd of psychoanalytic listening—a kind of concentrated antinoise, at such variance to the immediate environment—and it thrums through the heart of this Dallas night.

But what if there's nothing to hear? Carrying my salt-rimmed margarita, circulating, I feel an intense pang, a sense of massive alienation from all these psychoanalysts, from their Freudian belief systems—having been sent packing, so to speak, right after finishing residency. Like some defrocked priest back at the

Vatican, perhaps, or a clean-shaven Hasid back among his black-clad brethren, no longer observant of the 613 commandments, no longer chanting the daily prayers. They crowd around me, my former classmates, Jefferson and Annabelle (showing snapshots of her toddler), far less rebellious than I was, who remain at Payne Whitney; they're finishing their chief resident years while taking first-year analytic classes—on the inside track and heading toward becoming full-fledged psychoanalysts. McNeil and Zooey have already started their training analyses, and Georgie, who is slightly less befuddled these days, after several years of four-times-a-week psychoanalysis, has joined William Alanson White's psychoanalytic institute. Not to mention Bo, in handcrafted silver-toed cowboy boots and a black string tie with a turquoise clasp, who is just about to start his training downtown at NYU Psychoanalytic Institute.

My old colleagues, my training classmates, they gab, they gossip about the pros and cons of the various analytic institutes, and I don't have a clue who or what they're talking about. They *belong*, for sure, all of them: I'm out. Sure, I mostly hated it when I was there, I couldn't wait to leave, but now that I'm gone, I feel a bit of a pang that it will never be my world again. Exile is forever, as far as I can tell. *But there's no way I would have been able to fit in*, I tell myself. *The place was all wrong for me, that world was wrong for me, even after four years. I'd much rather be confused.*

The thing being, in this circus of the new psychiatric world, where do *I* belong?

The mariachi band falls silent, and our chairman starts bragging: new residents, new books, new object relations and Kohutian seminars, innovative new psychoanalytic programs. There are huge piles of shrimp and crab claws on ice and rivers of tequila to pour into salt-dazzled glasses.

Afterward, we get fifteen people together and take off, roaring toward Fort Worth. I'm driving, McNeil's beside me, wearing a ten-gallon cowboy hat, and three others are squeezed in back. Other cars follow. I think momentarily of my power analysis, as yet undone.

Soon we're at Joe T. Garcia's in Fort Worth, a crowd of fifteen half-drunk psychiatrists, being gawked at by the Texans on line, enough of us that they open up an outside seating area, pull several tables together under gnarled live oak trees, and bring four pitchers of margaritas right away and platters of enormous nachos—and four more pitchers—about five minutes later. My sense of alienation has faded.

I find myself thinking, hey, it's not so bad, screw the whole psychoanalysis business, in fact maybe this is the most exciting time ever to be a psychiatrist, look at the amazing new diagnostic system that doesn't require making assumptions about unconscious mental processes or transference or fantasies. New ideas that can be tested objectively and that have begun to spawn a new range of treatments that can also be tested rigorously. Understanding subtypes of depression, using antidepressant medicines for new conditions, eating disorders and obsessive-compulsive disorder and anxiety disorders. Developing brief evidence-based psychotherapy approaches, looking for measurable results in weeks rather than months or years.

Now that I'm working daily with it downtown at Manhattan Medical Center, it's clear that *DSM-III* is incredible: it's the first *practical* book psychiatry has ever seen. And the stuff I'm starting up, the work with patients, even the possibilities of doing research: it can really be exciting, and there are no limits on what we can do there, free from all the unwritten rules of the psychoanalytic life.

And if you think about it, we may have come from New York, but we *are* Dallas—*we will be the skyscrapers that replace the mansions. This is going to be our era, whatever comes of psychiatry in the next fifty years: in this new age of the* DSM, *it's up to us! But what about psychoanalysis? Is its time up? Or is there a way to make it central to our work in this new era?* I start talking a little about this, but fortunately nobody listens.

Then they all want to go off to Billy Bob's, which has two bands and five dance floors and nine bars and a shooting gallery and three stores and eight restaurants and an *indoor live rodeo*, so we drive off to Billy Bob's and drink an additional quantity of beer, and McNeil and Annabelle and Bo and I all take turns shooting it out with a mechanical cowboy for a quarter a pop, and then we start playing pool, and I'm making all kinds of impossible shots, angled, banked, Englished, off two or three balls into the side pocket, and nobody can believe that I can't play pool worth beans.

Which explains why the next day around ten a.m. I'm sitting massively hung over in the ten-mile-long lobby of the Anatole Hotel in my rumpled summer-weight Bloomingdale's suit, poring nervously over a volume titled *Power Analysis*, a large packet of charts and tables before me on the coffee table, and why even two cups of sludgy drug company coffee don't help me make much sense of what

I'm reading, and when I look up and see Jorge Luis Borges walking by I'm quite certain that I'm hallucinating.

But *it's him*, Borges, no question about it: the blind poet, noble visage, pale, an elderly white-haired man in a coal black suit, emerging from the elevator, led down the hall excruciatingly slowly by a pallid young woman with leaden-gray hair. Somehow, somebody convinced this giant to come here.

She guides Borges past the check-in desk to the gift shop, a crowd of psychiatrists in suits hurrying past them, in overdrive: researchers, professors, hungry-looking graduate fellows, and a few truants in jogging shorts heading the other way, toward the outdoor track. For a moment, it's as if we have all been imagined by him, by Borges, all of us characters in a story he is still trying to decide whether or not to write.

"Oh, *there* you are." It's Harold, a Payne Whitney professor who helped me with the research study I'm presenting in my poster. I'd sent him the revised text a few weeks ago. "Glad to run into you before your presentation. I had a few comments." He pulls it out of his blue vinyl APA valise and crouches next to me, showing me his edits. "Oh, how did the power analysis turn out?"

"I did it for suicidal and assaultive acts. Power was low."

He frowns, and my heart jumps. I look up. Led by his young woman, Borges rounds the corner, heading toward the atrium, then disappears. "Maybe I did it wrong. I mean, if you have a negative result . . ."

"You may have done it backward," he says. "Don't worry, probably nobody'll ask. Just fix it before you send the paper off."

I look at my watch: just twenty-five minutes to get down to the convention center and set up the poster. I hurry to my car and zoom downtown.

The poster display area is buzzing. All around, researchers are unrolling elaborate displays that they will mount on poster boards: computer generated, with four-color charts, photographs, scatter diagrams, professional lettering. I scrounge some thumbtacks and pull out my display. It's not much: a double-strike black-and-white printout from my word processor, blown up to five times its original size and printed on glossy photographic paper.

I tack everything up. Around me, the elaborate displays are taking shape. My project, "The Significance of Command Hallucinations," well, to be charitable, it looks low budget. Theirs are large single-printed glossy posterboards. Mine? A bunch of printed eight-by-eleven-inch pages thumbtacked to a cork background. Also, off-center. I rearrange the tables and abstract on the board, stand back, and try to convince myself that it doesn't look so terrible after all.

It's a weird thing that has begun to emerge in the midst of my training to be an analytic therapist—a researcher "self" has started come out. I'm not sure why exactly. Viewed psychodynamically, is it a manifestation of my "conflict" with my father and any authority figure? Am I acting out against the entire profession of psychoanalysis in some way? Or is it perhaps a "reaction formation," for example, instead of fighting with my Dad, I'm turning obedient, perversely complying with his wishes? Or is it a *true* interest in research? Which would be strangest of all!

Clearly it will require many hours of talk with Dr. V for me to resolve this issue.

It's noon. People start coming around, walking past the displays. "How'd you get such high numbers?" a man asks me. His nametag reads "National Institute of Mental Health." "Data on 789 patients—you just don't see that kind of numbers in psychiatry."

"Oh, we have a computerized database," I brag.

He reads my poster's conclusions. "That's good to know. We need more research like that." He pats me on the shoulder, then moves off.

I start feeling pretty good. Part of a noble medical tradition, the quantifying of clinical impressions, confirming or discarding them, an avenue to truth. I'm surrounded by high-science types in the poster room, by techno-MIT-engineer types who know how to use those fancy new multimillion-dollar machines to watch the brain at work. These guys hardly look like psychiatrists at all, at least the ones I've been surrounded by in New York, but more like nuclear physicists or organic chemists . . . heck, maybe my project's not so bad after all.

As a kid I was always fascinated by Dad's slides of abnormal EKGs and vectorcardiograms and scatter diagrams, and his articles in *Circulation*, profiles of him in *Time* and *AMA News*. This was *real*, this was what counted in the world: the advance of science, the conquest of disease. Though the Payne Whitney residency training program didn't teach us any research skills to speak of—statistics, research methodology, or the like—still, it's somehow important for me to come here, to show that I too can think like a scientist, can construct hypotheses, analyze findings, present my data to the world. And oddly, of my dozen classmates, I'm the only one who has any research to present.

A nattily dressed professor has stopped before my poster. He reads the abstract intently, then scowls. "What's this?" he says. "'The presence of command hallucinations is not correlated with suicidal or assaultive acts.' Did you measure the *degree* of commands?"

"No, just presence or absence."

"Then you're incorrect. 'Correlated' means you measured on a scale. You must mean 'associated.'"

"Oh," I say. I flush: he's found me out. "But the finding's still the same."

"What about for schizophrenics?"

"I . . . I didn't put up that table. But for the subpopulation of schizophrenics, there was no difference either."

"Mmmm." He nods, and I dread the next question, about statistical power, which will reveal how woefully lacking our research training was at Payne Whitney. We spent all our time learning psychoanalysis—how the hell would we have any time to learn to do research? Fortunately, he moves away without asking.

Finally, it's two p.m. I hurriedly take down my poster, buy a Klondike Bar for lunch, and rush into the enormous ballroom, standing room only, where Jorge Luis Borges is speaking on metaphor. There he is, way up front, speaking hesitantly in deeply accented English, swallowing many of his words, a near-blind man facing an invisible audience of thousands of dream interpreters and brain explorers that only he could have conjured.

He talks of the moon as the undying symbol of love. He talks of Helen of Troy, whose face launched a thousand ships. I'm reminded how, in *Ficciones*, Borges quotes the ode by Edgar Allan Poe in which the poet enters an enchanted garden where he sees his love, the incomparably lovely Helen; then, while he watches her, her visage begins to fade, so that finally only her eyes remain, glowing, to guide him through life.

Afterward, hundreds of psychiatrists swarm close, seeking the blind man's autograph. And of course McNeil appears, trailed by Bo, in his silver-tipped cowboy boots. "C'mon, let's get one," McNeil says. We push through the crowd, finally approaching close enough for a hug. The young woman with bullet-colored hair stares us down before we can snag Borges's scribble, and I manage to drag them away.

The next morning, pursued by Helen's invisible eyes, I struggle to become just another conventioneer in the enchanted garden, wandering restlessly from symposia to kiosks to posters.

A strange thought enters my mind and won't leave. *Are her eyes a metaphor, perhaps, for psychoanalysis, as it fades away, leaving its glowing vision?* I go back and forth with this, flip-flopping between scorn and admiration. I'm so happy to have escaped. And yet . . . maybe there *are* lasting benefits from the rigors of psychoanalytic training? Even if my patients at the clinic downtown aren't often good "analytic candidates"—they rarely are!—maybe something essential remains? Listening closely, with deep attention to our patients' spoken and unspoken words, or examining our own emotional reactions and trying to uncover our own biases, perhaps. Being a good doctor, an ethical and often self-effacing psychotherapist, listening deeply to our patients, not rushing to impose our own agendas. Yes, those things for sure. Being calm enough to allow the truth, however painful, to emerge from deep in our patients' minds. These are all principles of psychoanalysis, and all seem deeply relevant.

I feel torn at that moment, at once ill-prepared for modern psychiatry and at the same time rejected and abandoned by the psychoanalytic world. Somehow both badly and oddly well served by my training. And by my deep introspection itself, clearly marked by my analytic training. And my endless, endless ambivalence about everything, which has sprouted and mutated over my Payne Whitney years. Further, especially obvious here at a conference that shows the new directions of psychiatry is taking in the 1980s, I am almost aghast at how much time we have spent studying exclusively one tiny area of a huge, massively complicated world.

Later, at the drug company dessert party, I run into McNeil again. "Haven't seen anything to knock my socks off," he says.

"Me neither," I lie. It's too hard to explain, far easier just to agree.

Way after midnight, we wander back through the lobby of the Anatole, still packed with thousands of woozy psychiatric conventioneers. At this hour a casual observer would have a hard time telling whether we were trial lawyers or air conditioner salesmen or fuel-oil distributors. Then the APA president wanders past. Gray-bearded, gaunt as Lincoln, he sports a Stetson hat and a tuxedo, and he grasps a stogie, some strange archetype condensed into reality. He is one of my former Payne Whitney supervisors, a psychoanalyst, of course. Around his neck hangs an enormous medal with a portrait of Benjamin Rush, the very first American psychiatrist, from the 1700s, who signed the Declaration of Independence but was also an appalling racist, if I have it right. A Founding Father of us all, civilians and shrinks alike, for our ambivalent transference attachments.

Some residents rush by, Bo and Georgie and Zooey, on their way to the disco: Do we want to come?

"No, no, no!" McNeil says suddenly, and at first I think he's answering their question, but it is something else. The short, plump psychologist approaches. "I'm dead," he says, "She's got me."

"See you next year," I say and hurry for the elevator.

And then I'm in my skyscraper room, gazing at the streaming freeways and beyond, at the gleaming postmodern angles of a head-start on the twenty-first century. Practically everything here is jarringly new. And where do we—and I—fit in?

CHAPTER 4

Dreams of the Insane Help Greatly in Their Cure

Demolition of the Psychoanalytic Mothership, 1994

Then, I hear: *They are tearing Payne Whitney down.*

How is that possible? Isn't Payne Whitney an immortal monument not only to the psychoanalytic community but to New York City? And to the world?

My residency training at Payne Whitney lasted from 1980 to 1984. By this time, 1994, I've been gone for nearly a decade, and I rarely return to the Upper East Side. After four years of residency training at Payne Whitney Clinic, I left as an exile of sorts, an escapee, glad to leave the oppressive psychoanalytic world behind me. My refuge, Manhattan Medical Center, is a community hospital in lower Manhattan, where I've been working in the outpatient division, treating patients, doing studies, writing articles and books, and practicing a very different type of psychiatry than I learned during my residency years. So it's probably no surprise that I'm one of the last to hear of its demise.

I learn about it at an evening psychiatric meeting for the local district branch of the American Psychiatric Association: one of my fellow board members tells me that there's a garage sale of furniture, paintings, and other memorabilia, in case I want some. "Some real bargains," he says. "Early-twentieth-century Mission style furniture, signed prints by famous artists."

I ask him to back up: is it true? The whole building is coming down?

"Yes, maybe six months from now."

Souvenirs are the last thing I want. But over the following weeks, I can't get it out of my mind. And I'm not alone: by phone calls and emails and hallway conversations I realize that something is happening, as we Payne Whitneyites from around the city, loyalists and turncoats alike, all around the country, in our offices and institutes and clinics, gradually start to get the news.

In shock or glee, we phone each other, we pull each other aside in hospital corridors and at psychoanalytic institutes and after grand rounds. Yes, we know that all cities have their monuments and that some happily exist in the present for perpetuity. And that others, the unfortunate edifices, like the old Penn Station, which hovers ghostlike over West Thirty-Third Street at Seventh Avenue, exist only in memory. Still other corporeal monuments seem robustly immortal, only to be quickly shown vulnerable, stricken by fatal illness. We had never imagined that, before the end of the twentieth century, the Payne Whitney Clinic could be such a place.

The crowd I came up with—psychiatrists spanning the Age of the Couch and the beginnings of the Age of the Clinic—comprised only a small fraction of the acolytes of this monument. For sixty years, psychiatrists and patients and millions of other Americans, I soon come to realize, knew the Payne Whitney Clinic as an enormous white-brick transference object located on Manhattan's East Side, hard by the East River. Payne Whitney was a massive touchstone in Manhattan's psyche. It was a center of stability in a stressed-out city, a seemingly immortal cathedral of the psyche.

I had long hoped for and dreamed of this moment. I wanted Payne Whitney to live forever and at the same time to be destroyed immediately, smashed into dust. Especially when I worked there. And in this latter, often-fervent wish I doubt I was alone: I am certain that many fellow doctors, not to mention innumerable patients, had dreamed of the same thing since the very moment it was built. How many times had I cast curses upon Payne Whitney during my own dark hours, especially early in residency training? And hadn't hundreds or thousands of New Yorkers, incarcerated in its drafty wards, prayed for a thunderbolt from the heavens?

Over the coming weeks, I begin to imagine Payne Whitney gone, destroyed, replaced, and perhaps forgotten. It is a strangely upsetting prospect to me, even painful, and I'm not clear exactly why. Plenty of old buildings come down and no one really cares, so why is this one shaking me up even in advance?

I begin to steal time from my home and work lives and start visiting the medical library at Manhattan Medical Center, which has a few medical history books, quickly exhausting that resource; I talk my way into the New York Hospital library, which has an extensive medical history collection. And I drop by the local public library near my home in Westchester County for more popular works. When our local library doesn't have the items I want, I begin to order through interlibrary loan. In the summer of 1994, I even finagle a journalistic assignment (from the *New York Times*, no less) to visit Payne Whitney, to write an article for the *New York Times Magazine*, a pretext for a chance to say goodbye.

Upon which I begin my researches in earnest. What do I find? That before becoming a building and a myth, Payne Whitney had been a man. A reclusive, immensely wealthy early-twentieth-century financier and thoroughbred horse breeder, the second son of William C. Whitney, one of the richest men in America, William Payne Whitney was educated at the Groton School in Connecticut, attended Yale University and Harvard Law School, and was a member of the exclusive club Skull and Bones. Following his untimely death in 1927 at the age of fifty-one from an attack of "indigestion" while playing tennis at Greentree, his Manhasset, Long Island, estate, Whitney's will provided twenty million dollars from his $100 million estate to build a tall white-brick clad complex, the New York Hospital. Specifically, it instructed that the new hospital include a separate psychiatric clinic named for him.

Those of us who trained there in the 1980s joined previous generations of psychiatrists in wondering why on earth this aristocratic scion would have wanted his name connected to the psychiatric clinic and why, of all the divisions of the vast hospital center, his portrait hung in *our* building's lobby. Lacking any definitive answer, we speculated that it must reflect some secret suffering of the tycoon himself or of someone he loved: his wife, an ill child, or perhaps a mistress.

Whatever his motivation—upon which my haphazard new researches shed no new light—Whitney's enormous bequest was reported to have given both glamour and an air of legitimacy among Manhattan sophisticates to what had been a marginal and secretive profession. Those of us who trained there inculcated its glamour into our very bearing and carriage, by the way we wore our priestly long white coats as we strutted through its hallways, speaking our strange psychoanalytic patois.

But where does Payne Whitney Clinic fit in the history of psychiatry? I began to read about the history of madness and its treatment, how it was largely grim and shameful, long lagging behind the care of the medically ill. While traditionally in Europe the mad were kept at home, cared for by their families, by the seventeenth and eighteenth centuries, private and public madhouses had been set up, and through the 1800s American states proceeded with the development of large hospitals that provided little more than custodial care. In the later nineteenth century, in the context of massive immigration from Europe, state mental hospitals began expanding rapidly, so that by 1904 the United States had relegated over 150,000 people to life in asylums.

What could be done with the growing masses of the mentally ill? Were they doomed to lifelong incarceration, to endless suffering?

In 1897 came a dramatic discovery by the Austro-German psychiatrist Richard von Krafft-Ebing (best known for his later treatise *Psychopathalia Sexualis*, about sexual pathologies): neurosyphilis—an infection of the brain by the spirochete *Treponema pallidum*—was the cause of many cases of madness. Soon, new experimental treatments began to be introduced to psychiatry. In 1917, "fever therapy" was introduced to treat neurosyphilis—patients would be deliberately infected with the malarial parasite *Plasmodium* to treat their psychosis, somehow curing one infection by deliberately causing a second one, fighting fire with fire. In 1920, barbiturate-induced "deep sleep therapy" was first used to treat what was then called dementia praecox and is now known as schizophrenia. By 1933, insulin shock therapy was introduced, and soon after that, in 1935, psychosurgery, followed by the gruesome prefrontal lobotomy procedure in the late 1940s.

Over the same period, there was a growing appreciation for psychological factors as the causes of mental suffering: in the late 1800s, the psychologist Pierre Janet and the neurologist Sigmund Freud were describing hysteria as arising from split-off memories of traumatic events. The onset of World War I led to the development of thousands of cases of "shell shock," which affected as many as 40 percent of combat casualties and required the development of new forms of treatment—including the most effective approach, behavioral exposure, in which soldiers with shell shock were quickly returned to combat rather than being withdrawn to remote hospitals for treatment.

In 1909, Freud came to America and gave five lectures at Clark University in Worcester, Massachusetts. Soon many American psychiatrists began applying Freud's new psychoanalytic methods to treat mental illnesses. In 1913, a full-page *New York Times* article, headlined "Dreams of the Insane Help Greatly in Their

Cure," described how "theories of Dr. Freud are put to practical use at Ward's Island and in other institutions in the City that care for the mentally disturbed":

> There has been a revolution in the last few years in the method of treating the insane and those persons suffering from nervous afflictions bordering on insanity. The new mode of procedure is based on the remarkable studies of Prof. Dr. Sigmund Freud of Zurich [*sic*], who has devoted his energies to a field of investigation practically untouched by other research workers engaged in elucidating the mysteries of the disordered mind.
>
> Freud's method is known as psychoanalysis, which bears the same relation to mental and nervous diseases that the microscope does to pathology. Abnormal mental conditions had been judged hitherto practically by mere superficial inquiry and observation. Freud and his pupils literally turn the minds of their patients inside out.
>
> (*New York Times*, March 2, 1913)

The early twentieth century was a time of much ferment and hope in psychiatry. Could that have influenced Payne Whitney's bequest?

This is where my researches for the *New York Times* article, desultory as they were, came to a dead end. I could find no record of Payne Whitney's inner life, of his personal history, no evidence of diaries, letters, journals, any autobiography or biography that could shed further light on our benefactor's motivations. His 1927 *New York Times* obituary describes in detail a passion for horseracing and generous previous gifts to New York Hospital and the Bloomingdale Asylum, located in White Plains, New York. Biographies of his daughter, Joan Whitney Payson, focus disappointingly on her ownership of the New York Mets, her love of thoroughbred racing, and her estimable collection of impressionist and post-impressionist paintings, which were donated to the Metropolitan Museum of Art upon her death.

What I am left with, alas, is a mass of ill-grounded hypotheses, rampant speculations, and projections of my own frazzled mind and those of my colleagues from one decade to the next. The best I can come up with for an origin story of Payne Whitney Clinic, in the mind of William Payne Whitney the man, is something condensed, precipitated, from the American atmosphere of the 1920s, from the culture itself, with the force of a vision, a vision inducing a psychic readiness in an unlikely sportsman and titan of society: a strange enthusiasm for building a psychiatric hospital adjacent to a great medical center, as a part of the city.

A compelling vision—even if analyzing the dreams of the insane did not actually cure insanity but was better suited for the treatment of neurosis.

By the mid-1930s, I now read, Payne Whitney Clinic had become the place New Yorkers went to repair their psyches. A place to recuperate, to dry out, to regain sanity. The clinic was the genteel antipodes to the frightening chaos of Bellevue. It was not a place of cutting-edge science or theoretical investigation but a humane place for healing based on the idea of psychoanalytic psychotherapy. Payne Whitney represented order and hope—from its gracious architecture to its highly organized inpatient units.

I find this part of my reading and research particularly evocative: how the building was designed like an elegant hotel, its marble-floored lobby graced by comfortable furniture and an ornate fireplace. At the same time, even more impressive, the Payne Whitney building was also engineered for madness, with special hydrotherapy tanks, steel-framed casement windows that opened only five inches to prevent suicide, and soundproofed ceilings. Severely agitated patients could be calmed by the latest 1930s technology, such as "continuous infusion baths" or by "wet packs," cold-water-drenched towels that slowed the wildest metabolism—but from Payne Whitney's beginning, the mainstay of treatment was psychotherapy. (Incidentally, though cold baths and cold packs were debunked for many years, recent research shows that cold infusions and baths stimulate the vagus nerve, increasing the activity of the parasympathetic nervous system. Vagal nerve stimulation is now an established treatment of depression.)

And there was no shortage of Americans seeking psychotherapy. Starting in the early 1930s, innumerable wealthy New Yorkers began to check into Payne Whitney Clinic for months at a time, even for a year or more, seeking psychic solace. As psychiatry residents in the 1970s and '80s, we would find old letters from illustrious Manhattan families tucked into the backs of heavy wooden desk drawers, beautiful penmanship on high-rag-content stationary addressed to incarcerated relatives, letters we would shove back deep into the drawers, keeping them secret forever, or at least until discovery by a future doctor-in-training.

But, perhaps surprisingly, the Payne Whitney Clinic soon took its more-or-less-illustrious place in American culture—inspiring some of the greatest artists of the day. Now I discover how the great poet Robert Lowell, in a manic frenzy, once stared through the same metal-framed windows we psychiatry residents peered out of, desperately searching for a road back to sanity, whereas we were just seeking courage to deal with meeting our next supervisor. Pulpy tell-all biographies, accessed in a quiet New Rochelle Library reading room, tell me that the great Marilyn Monroe stripped herself naked, screaming, in a Payne Whitney seclusion room.

Lowell was one of the first Payne Whitney patients treated with the drug Thorazine, which was synthesized by French scientists in 1949. Though he was tormented by side effects ("my blood became like melted lead"), a precarious sanity returned, and he was discharged after five months. His stay at Payne Whitney marked a change in his poetry, which became increasingly autobiographical and confessional, culminating in the 1959 publication of his classic book *Life Studies.* In his poem "Man and Wife" he writes how his wife, Elizabeth Hardwick, "faced the kingdom of the mad—its hackneyed speech, its homicidal eye—and dragged me home alive." In a sense, hospitalization did the same for him. For the rest of his life, though plagued by attacks of mania and depression, Lowell would be helped by the newly introduced drug lithium.

It was clear from the celebrity biographies—and piles of old *Life* magazines—that Payne Whitney's most famous patient of all time was Marilyn Monroe. She came in 1961, after divorcing the playwright Arthur Miller. For Monroe, Payne Whitney was a torture chamber. "They put me in a cell (I mean cement blocks and all) for very disturbed, depressed patients," she wrote to her psychiatrist, "except I felt I was in some kind of prison for a crime I hadn't committed."

After three days of what was reported to be high drama, the baseball immortal Joe DiMaggio flew from Florida to get her out of Payne Whitney. Monroe was then admitted to Columbia-Presbyterian Medical Center, in far uptown Manhattan, in the shadow of the George Washington Bridge. Neither hospital, of course, was able to modify her downward spiral, and Monroe overdosed barely eighteen months later.

(I hasten to assure the reader that when we trained there as psychiatric residents, we were only vaguely aware of Payne Whitney's history. Our professors might or might not have treated the celebrities whose biographies provided scandalous details; they kept any such confidences utterly intact, and consequently it was only from our outside readings that we learned the tales, true or not, behind its mystique.)

But for those of us who worked there, we were far more aware of the thousands of patients untouched by celebrity, for whom the Payne Whitney Clinic played an equal number of roles. Some were soothed, it seemed, by its protective environment, others rightly repelled by its lack of physical amenities. Still others complained of feeling suffocated by its inflexible rules.

The ultimate New York transference object, Payne Whitney became haunted by ghosts famous and obscure.

I conclude this during my research: if there is a secret to Payne Whitney's success over the decades (and the clinic's gloomy hallways and consulting rooms have held many a secret), it resides in the idea of finding asylum in the city. Payne Whitney Clinic was constructed to be physically connected to the rest of New York Hospital yet separate from it. Perhaps because of the ancient fear of madness, or perhaps from early-twentieth-century psychiatrists' insistence on autonomy, the clinic was built as a small self-contained asylum on the grounds of a general hospital, linked to the rest of the complex only by underground tunnels.

Even half a century later, this connection was compelling—the thought of learning to be a psychiatrist at a psychiatric hospital located close by a massive hospital or of being a patient in a psychiatric setting close by a medical facility in case things go south. (Payne Whitney Clinic was not alone for its era: in the same decade, the New York State Psychiatric Institute was also built adjacent to Presbyterian Hospital. No doubt there were many such hospitals built around the same time: psychiatric asylums constructed on the campus of general medical hospitals, a type of institution whose history has not yet been written.) In contrast, the idea of doing our training at an old rural asylum that had been converted into a modern psychiatric hospital—and there were illustrious examples, like the McLean Hospital outside of Boston and the Institute for Living near Hartford, Connecticut—seemed somehow old-fashioned, even retrograde.

Just as Payne Whitney Clinic was a touchstone for patients, so too was it an emotional center for generations of New York psychiatrists, especially after the arrival of the psychoanalyst Dr. Robert Michels in 1974, who served as chairman of psychiatry for nearly two decades. Michels brought psychoanalytic rigor to what had gradually become an anachronistic hospital catering to the rich. He strengthened the connection of psychiatry to medical care. The expanded consultation liaison psychiatry service provided psychiatric care for patients in the medical and surgical floors of New York Hospital, including the kidney transplant service and the burn unit, proving how valuable psychiatric care could be for people with serious medical illnesses. He also enhanced clinical and research programs—but most of all, he improved residency education, increasing the number of residents from twenty-six to forty-five and introducing a rigorous new psychoanalytic therapy curriculum.

Soon—a few years before I applied to train there—Payne Whitney became the most desirable psychiatric training program in America. Dr. Michels was a flamboyant psychoanalyst, if such a thing is possible. He was notorious for stopping residents a few sentences into a case presentation and tearing them to shreds. He would stand up in departmental grand rounds every week and upstage the speaker by asking three (always three) devastating questions. We mocked him in our annual Christmas show to the tune of "Jesus Christ Superstar" (even crucifying the actor playing him in the climactic scene), yet we aped his psychiatric style as we strode the shadowy hallways in our starched white coats and as we dispensed Michels-inspired pithy interpretations.

In the Bob Michels era, residency training was psychic boot camp. We residents gossiped that the Payne Whitney educational method entailed a deliberate attempt to *induce* deep psychological crisis (and then a referral for psychoanalysis) among those of us in training, its aim being to make us anew. Certainly, that's what happened for me. Every one of my classmates, Bo and Georgie and McNeil, and Jefferson and Annabelle (our former chief residents), and the rest, it seemed, was in psychotherapy, preferably in full-fledged psychoanalysis. After four years, for many of us the longest four years of our lives, we emerged having inculcated the famous Payne Whitney strut, a combination of arrogance, empathy, and a deep conviction that we alone knew what was best for a patient. To us, Payne Whitney embodied psychiatry as it should be and always would be.

By the early 1990s, though, we begin to encounter former professors at conferences and meetings, their hauteur strangely shaken. We hear that well-known faculty members are leaving for better opportunities elsewhere. Budgetary problems loom, and bed occupancy drops, forcing a decrease in inpatient beds from 104 to seventy-two and staff layoffs. Senior physicians retire; many are not replaced. Michels became dean of Cornell's medical school in 1991, and when I run into

Payne Whitneyites, they complain of low morale and a leadership gap. Then comes the plan for a *new* New York Hospital. We hear that the president of New York Hospital, a surgeon, is fiercely determined to build a new building and has his eye on the Payne Whitney site. According to gossip, at the crucial moment of decision, the Psychiatry Department lacks the will, or the clout, to oppose him. By the time a new chairman, Dr. Jack Barchas, arrives in the fall of 1993, the demolition of the Payne Whitney Clinic building has been confirmed.

And indeed, demolition is imminent at the time of my visit for the *Times* article. First, I meet with the architect, a middle-aged woman with gray shoulder-length hair who is toting a large bundle of floppy blueprints; we ride to her office high up in the main New York Hospital tower, and she unrolls them before me, showing me the plans for the new building and how it will cantilever over the FDR Drive. Every effort was made to save the old building, she tells me, but in the end, its ceiling heights were too low for modern hospital requirements: nothing could be done.

Then I meet with Dr. Michels, in his dean's office at the medical school. I have always been in awe of him, and we both seem aware of the irony that I, as a bit of an outcast and rebel, someone who long ago rejected the idea of becoming a psychoanalyst, am conducting this interview. He tells me his view of the trajectory of Payne Whitney and its importance in the history of psychiatry.

I want to challenge Dr. Michels that day about the model we spent so much time learning and about its hidden costs: *But what about all the things we didn't learn?* We learned almost nothing about research methods, or statistics, or cognitive behavioral therapy. We spent so much time on psychoanalysis: *why?* Especially since we still don't know if it works, why put so many resources into it? I don't ask these questions, though, which I regret the moment I leave his office. (Ironically, when the article is finalized for publication, Dr. Michels's comments are cut—according to my editor, "for space reasons.")

Finally, I walk to the temporary Psychiatry Department floor to meet with the new chairman, Dr. Barchas. Unsurprisingly, he is not a psychoanalyst. No one hires psychoanalysts to be chairs of psychiatry anymore. "It's a transition point in the discipline," admits Barchas. A balding, avuncular man with dark-rimmed glasses, whose quiet attentiveness is oddly reminiscent of a psychoanalyst's, he is instead known for basic biological work on neuroregulators in the brain.

What are his goals at Payne Whitney? I ask, imagining that's what a real journalist would want to know.

He gives a thoughtful answer about science, neuroimaging, and genetics and neurobiology and how he wants to boost psychotherapy training, even bringing in the internationally known psychoanalyst Otto Kernberg to start a new psychotherapy center.

It is a terrific answer, but at that moment—a gatecrasher at a funeral—I almost pity Barchas. He will soon be running a homeless department, wandering and waiting until its new residence is completed five long years from now.

So, yes, Psychiatry—the inpatient units, that is—will return and occupy the new building's top floor. Again, Payne Whitney will gloriously face out over the East River's rushing tides. "We will have sixty beds, the entire top floor of a building almost two blocks long, beautifully set up for nurses and patients," Barchas expounds that day. Inpatient psychiatric care can be better integrated with other medical specialties. An improvement, since psychiatry is now dealing with very ill patients, many having medical problems.

But that is the least of it. As we talk, the new chairman and me, I realize that the unthinkable is happening at Payne Whitney: not just the demolition of a shrine but the open admission by its own leader that psychoanalysis is no longer the apotheosis of therapies but only one of many approaches. Surely, we knew this from the very moment that the *DSM-III* arrived over a decade ago. But now, the Vandals are readying their wrecking ball.

Indeed, what to make of the legacy of the psychoanalytic era? As we gab, and afterward as I try to figure out what to put in my article, my mind veers away to larger issues. What have the psychoanalytic perspective and model brought to psychiatry over the past half century? Having been away from Payne Whitney for a decade, I am now seeing it afresh, more critically, also perhaps more sympathetically.

If I have to summarize, to expand on the thoughts I had while interviewing Dr. Michels and Dr. Barchas, I would say that psychoanalysis, at least the classic four-or-five-times-a-week treatment, is fantastically impractical, not only because of its time commitment and cost but also because it can only help a small fraction of patients. It does seem to help people with neuroses and with some personality disorders, but the data for "efficacy" is extremely limited. Having struggled to start my own psychotherapy and psychopharmacology research studies, I keep being amazed at how few studies of psychoanalysis have been done since it began to rule psychiatry. Research is just not central to that treatment approach, for one thing; theory is central. Because of that, it's incredibly difficult to capture the essence of psychoanalytic treatment in designing research studies, much more difficult than studying medication treatment or brief treatment with cognitive behavioral therapy. So that's not what sticks with me, after a decade away from Payne Whitney.

In my Payne Whitney days, I was often aware of the limits of psychoanalysis. Its rigidity, the unfalsifiability of many of its theories—you couldn't do a study to disprove them—plus the use of interpretations as a means of social control in the tiny hothouse world of psychiatric training and the almost comical arrogance of many analysts. All these things seemed antithetical to the original spirit of psychoanalysis but highly prevalent in psychoanalytic culture for reasons that were not too difficult to understand. They seemed strangely analogous to the rigidity of the Catholic Church in contrast to the radical acts and words of Jesus. Perhaps in leaving Payne Whitney, in abandoning the psychoanalytic enterprise, was why in my new life downtown at Manhattan Medical Center I kept feeling somewhat like a defrocked priest, an apostate of sorts.

From the viewpoint of public health, which I can't avoid from my work at MMC, there's a huge problem with how broadly applicable the psychoanalytic model is. So many patients that I encounter every day at MMC are "not good candidates" for psychoanalytic psychotherapy. They don't have enough "reflective capacity" or patience or time or money for several-times-per-week treatment for many years—it's time intensive and thus perhaps inherently "posh," feasible for only a tiny proportion of the population. Data then becoming available show that psychoanalytic approaches don't do much for the serious illnesses of bipolar disorder, schizophrenia, or obsessive-compulsive disorder. Lying on a couch, free-associating for hour after hour, is not only highly unlikely to help these conditions—it could make them worse. And after more than half a century, it's not clear if analysis alleviates conditions like major depression or panic disorder.

How to help all those who aren't willing or interested or able? Are they irredeemable? And in the end, what's the good of a model that only works for a fraction of cases? And at MMC, even if a larger percentage of our patients *were* interested and motivated to start psychoanalysis, we obviously don't have the staff or time to treat them. And who would pay for their care?

And yet, I realized as I sat talking with Dr. Michels, after weeks of reading about the history of Payne Whitney, I have not truly abandoned what I learned. In fact, I have been profoundly influenced by my four challenging and often torturous years there.

Those aspects of psychoanalysis that continue to help me as a psychiatrist in treating real-world patients downtown at Manhattan Medical Center? There are many, I must concede. At its best, the psychoanalytic treatment we were taught was deeply humane and respectful, in that it prioritized the creation of an intimate, ethical interpersonal connection between human beings. My training makes me pay attention to the importance of the therapeutic bond between patient and doctor and to the emergence of "transference" feelings, in which patients bring emotions and ways of relating from their previous relationships, especially early-life relationships, to their current life and to the therapy itself. The analytic mindset leads me to try to understand how those past relationships can influence current-day relationships with partners and friends and to try to use the patient-therapist relationship to create change in the present day. And the psychoanalytic idea of "interpretations"—of making observations about the relationship between patient and therapist in the context of larger life issues—I find entirely inescapable in practicing any form of psychotherapy.

A few other things stay with me: while I don't like the concept of "the unconscious mind" as a *unified* thing in itself, I can't help but be impressed by the importance of nonconscious processes that often rule our behaviors and the humbling ways in which we human beings are hardly the deliberate drivers of our own destinies.

Then too, I am increasingly struck by the importance of trauma, especially early trauma, throughout adult life, which I see time after time in my patients at Manhattan Medical Center. Whether Freud was wrong in concluding that his analytic patients often imagined their own abuse—and whether he did in fact conclude this is a matter of vigorous debate among historians—doesn't interest me as much as the fact that he focused on the *centrality* of traumatic experiences and their lasting impact.

And finally, another important lesson of psychoanalysis is the recognition of the importance of sexuality in all aspects of life. How much has psychoanalysis played a central role in this realization, in comparison to other movements, say, political or artistic? I leave this to the intellectual historians. But I find it difficult to believe that the method of free association and nonjudgmental exploration of mental processes didn't play a significant role. (Surprisingly, young gay and lesbian people now embrace psychoanalytic psychotherapy, despite the checkered history of psychoanalysis in pathologizing homosexuality; they are enthusiastic about using its approach to alleviate their suffering, to reconstruct their selves.)

And one last thing: the importance of psychological "attachment," which is central to much psychoanalytic theory. The stability or instability of connections between human beings is obviously so important to our quality of life, and people with "insecure attachment" often flounder in their efforts to build good relationships. In comparison, the *DSM*, which I now spend my days at MMC implementing with innumerable patients—doesn't even have a language for this. But it affects so many of my patients, especially those with "personality disorders" and those with histories of significant trauma. How can they trust others when they have been harshly betrayed, abused, or abandoned? How can they learn to make stable connections with others? Psychoanalytic practitioners and theorists have at least provided a language for this problem, even if it lacks clearly effective treatment approaches that would be applicable to the thousands of people we encounter who have this kind of problem.

By the fall of 1994, I come for a final visit, just before my *New York Times Magazine* article is to appear (to a chorus of calls and grieving letters). Demolition is well underway. The building will be entirely gone by December, I am told, and on its site will rise the *new* New York Hospital, a $900 million, 774-bed, 810,000-square-foot megalith—though the PR office would describe it differently—which will occupy a massive platform above the FDR Drive. This is hardly the end of psychiatry at New York Hospital, everyone assures me: outpatient psychiatric services have been relocated to East Sixty-First Street, and inpatient units are camping temporarily at another hospital on East Seventy-Sixth Street, before eventually coming back home to a space in the as-yet-nonexistent new building.

But optimism only goes so far: clearly, this is the end of an era. Standing there on a windy Friday afternoon, gaping at scaffolding, construction barriers, a blindly windowless façade, I find myself trying to recall the interior, to recreate it in my mind, to remember what was on each of its eight floors. I recall the long rows of therapy offices in the Payne Whitney basement and the much more posh first floor, with its paneled lobby and the chairman's and the other administrative high-ceilinged offices. I recall the third-floor depression unit and the long-term borderline unit on the fourth floor, where patients would stay for up to a year. But already I have forgotten where the geriatric patients were housed, and where the psychotic unit was, and the massive psychoanalytic library. The building's wooden paneling and ornate fireplace may be in storage, awaiting their reinstallation high up in the new hospital building, but the time for mourning has arrived.

The name "Payne Whitney Clinic" is now attached forlornly to awnings in front of temporary inpatient units on East Seventh-Sixth Street and an outpatient department on East Sixty-First Street, in a forgettable stone-and-brick office building buried in Manhattan's unforgiving grid. Vans continuously circulate between the department's several scattered sites, transporting charts and people, aswirl in constant motion, a periphery without center.

So gradually, as with any loss of a beloved object, the acute phase of mourning begins to pass, as we enter the stages of denial, protest, grief, and sadness, and Payne Whitney transports itself from our real city to the ghostly simulacrum of its memories.

Perhaps all too quickly, New Yorkers accommodate to Payne Whitney's disappearance, "sealing over" the trauma of its loss. And so the old Payne Whitney building has become part of a vanished city, along with Ebbets Field, Penn Station, and a thousand other demolished New York landmarks, recreated only in the timeless unconsciousness of memories and dreams or, in some cases, nightmares.

Perhaps, I conclude, by the end of the twentieth century the city no longer needs psychiatry in a separate house, confined to an asylum that is kept at arm's length from the rest of medicine and society. In the early 1900s, psychiatry had been largely an inpatient specialty, foreign and frightening; now it has become largely outpatient, practiced in the community. While Payne Whitney's inpatients will be indeed eventually be integrated with medical floors in the new New York Hospital building, the outpatient services have vanished into the city, which has thousands of therapy offices, places of momentary asylum.

And the psychoanalytic model, which Payne Whitney embodied, what has become of that? Distilled to its essence—since almost no one can afford full psychoanalysis, after all—it has morphed into "psychodynamic psychotherapy," an approach to therapy that requires less frequent visits, with the patient sitting in a chair rather than lying on a couch, and is far less costly, far more flexible, and less grueling. And its acolytes are everywhere. (But is it too watered down to be effective? I imagine my old supervisors, full of doubts.)

Indeed, all throughout Manhattan, in office and apartment buildings, one notes the suites and offices of psychotherapists, many of them Payne Whitney graduates, mine included—tiny gracious outposts of Payne Whitney, colonies of the now-annihilated mothership.

PART II

The Clinic, 1985–2000

CHAPTER 5

Treating the City

DSM Psychiatry in the Real World of the City Hospital, 1989

STRIKE DUTY

So far things are quiet.

I sit in our walk-in clinic, finishing paperwork on the patient I just saw, a lady from Avenue D. It's a blazingly hot day in July, and Mrs. Rodrigues is unable to stop weeping. Her son, her only son, is dying. He can't breathe, he has SIDA, she says. AIDS, probably pneumocystis pneumonia, incurable.

Or so I gather, since she only speaks a few words of English, and my Spanish is sorely lacking. Nothing can console her. Now she sits in the interview room, a Dominican woman in her early forties, moaning and wailing. What can I do? There are no bilingual therapists available to see her. I can't even assess whether she's suicidal and in need of inpatient hospitalization.

At a certain point, I sense something going on—a subliminal difference in the air, a change in the buzz of conversation from out front, a collective intake of breath. At the registration desk I find Franklin, our administrator, and Deb Silver, MD, our child psychiatrist, choking, barely able to breathe.

"I shouldn't have buzzed him in," Deb gasps. "He said he needed to use the bathroom, and I felt sorry for him."

"What? Buzzed who in?" I say.

Then it hits. Rancid and cloying, a horrible stink, reminiscent of what Klaus and I used to concoct back in fifth grade, a brew of chemistry-set sulfur plus dog shit twisted in brown paper then set afire. Tears spring to my eyes.

"This guy, I'd never seen him before, he went into the bathroom and set off a stink bomb," says Debra. "I started smelling it after he left. It's not so bad at first but after a while . . . God! I think I'm going to throw up!"

She flees down the hall toward the staff bathroom. I rush around the waiting room, opening windows, and Franklin phones Maintenance.

"What are *they* gonna do?" I say. "Maintenance is on the picket lines too."

"*Someone*'ll be there," says Franklin.

This is the first indication that things might get nasty, on this steamy day. I've been covering Manhattan Medical Center's Walk-In clinic, seeing the few patients who—ignorant of New York City politics or desperate enough to cross picket lines—have struggled through the police barricades and onto our floor. Once inside the building, they find their regular doc is gone, pulled to strike duty on the inpatient service. They're upstairs on the locked units, changing linens, transporting patients, pushing pills from the nurses' carts.

Basically, working as scabs, as am I. And the receptionists and clerks who usually greet them are gone as well—probably holding a sign outside. Hospital workers, Union 1199 members, have walked out. It's about money, of course. All that is left today is a skeleton crew. Besides Franklin, who's manning the registration desk, and Deb for child cases, there's me and two other docs for adult. Compared to the fifty staff usually on the floor. Downstairs, security is beefed up, and our lobby doors are locked, so all patients get buzzed in.

In a few minutes the supervisor arrives, a freckled, affronted-looking man in a blue hospital jumpsuit, brandishing a mop and followed by a thirteen-year-old black kid whose large nametag reads "VOLUNTEER," who wheels a thirty-gallon drum before him. For the rest of the afternoon, the kid sloshes various cleaning solutions around the bathroom and the waiting room. A vile, stinging mélange of ammonia, industrial solvents, and God knows what else mingles with the stink bomb's odors, until I retreat to my office.

There I open the windows and turn the AC on high. I see in the park across the avenue a knot of protesters, chanting and singing. They wear white-and-blue Union 1199 hats, and some carry picket signs. Whatever they are chanting, it is half-hearted, lacking in brio. No surprise, really. The main action is at Presbyterian Hospital, uptown, near the George Washington Bridge. Huge crowds, TV cameras, all that. We're a political backwater here, a place for minor confrontations, token acts of sabotage. Our stink bomb won't make the news.

MANHATTAN MEDICAL CENTER

By the summer of 1989 I've been at Manhattan Medical Center, in downtown Manhattan, for five years, in what has been a dream job. A big distance from my Payne Whitney training days: rather than haughty psychoanalysts, we're roll-up-your-sleeves community psychiatrists. And face it, the psychoanalytic treatment model I spent four years learning at Payne Whitney has almost no direct relevance here. Whereas the new *DSM-III* is proving enormously helpful in allowing us to make diagnoses and decide what treatments to provide.

MMC is plopped in the midst of the city. Sure, there's a park out front, but it's filled with junkies on one side and sullen high school kids on the other. We have patients from everywhere—the Dominican Republic, Puerto Rico, Haiti, Mexico, Brazil, and that's just the Western Hemisphere. They come from China, Korea, the Philippines, from India, Vietnam, Cambodia, Russia, Ukraine, Hungary, Yugoslavia—from across the world. MMC is in the city and *of* the city. We lunch at Lower East Side restaurants serving tapas, pierogis, hot pot, jerk chicken, Korean barbecue, kosher pastrami, sushi, Shanghai soup dumplings, and any kind of burgers you can imagine.

Every day, my fellow doctors and therapists are treating every known kind of psychiatric disorder and, no doubt, many unknown kinds.

You could say we treat the city, or really, the world.

Despite its impressive number of inpatient beds (nearly a thousand), MMC has a feeling of intimacy and friendliness, an utter lack of hauteur. Physically it is not particularly impressive, as it consists of a series of small buildings constructed over the past century, with linoleum floors, yellowed paneling, scuffed marble wainscoting, everything connected by a warren of zigzagging tunnels. Everything at MMC is named for someone—hallways, stairways, conference rooms—every surface covered by an engraving or plaque naming a dead benefactor. Should you read the hospital's early reports, you will learn about donations of chickens, gallons of wine, bequests of two dollars from the estates of grateful spinsters. Though its annual benefit dinner fills the ballroom of the Waldorf or the Pierre, you can still run into the hospital president on a street corner and chat for a moment, and he might even ask after members of your family.

MMC is not immune from the city's ills. One of our doctors, Awkete, a gentle immigrant from Ghana, got hepatitis when she was working on call last winter, exhausted, and jabbed herself with a needle she had just used to draw blood. Thom Singletary, one of our psychiatrists, a good friend, is losing weight and won't say why. We suspect he has contracted AIDS. Even in our own clinic, we may not be safe: One time a patient jumped out of his chair and began strangling his therapist. We heard her pounding the walls and pried him off her just in time.

On a more mundane level, things tend to disappear here. A sandwich left on your desk in an unlocked office, purses and wallets, of course, even a blue Acropolis cup half-full of lukewarm sweet coffee. Medication samples vanish from locked cabinets, printers, entire computers, and foldable wheeled stretchers are suddenly nowhere to be found. The city infiltrates us. I've come upon men shooting up in the bathroom stalls downstairs and homeless guys trying to wheel shopping carts in past the security guards. Our regular Walk-In doctor routinely asks patients what weapons they are carrying and tells them to put them on the desk in front of him. Guns, knives, box cutters, ice picks, you name it.

Yet I love it here. Not only is Manhattan Medical Center booming with new energy, but so are our surroundings, with a host of new immigrants, some of them refugees, others yuppies, artists, gay people. We see Keith Haring plastering posters on the subway, pass by the Palladium discotheque, and see Jean-Michel Basquiat's work in Lower East Side galleries. They are accompanied by new catastrophes. AIDS, crack, crime, homelessness: so many new epidemics are surging in our part of town, with hotspots just a few blocks from us, and we here at MMC are trying to take them on.

And MMC is an ideal place to treat the city. Our psychiatry department has been growing rapidly, as has the rest of our hospital. To respond to the city's needs, we've been setting up all kinds of innovative programs, ones that use the latest advances from psychiatric research and incorporate them into the clinic. Psychiatry is in transition too. Just a year ago Prozac arrived, the first of the new selective serotonin reuptake inhibitor meds, which are certain to revolutionize psychiatry. Say all you want about the evil drug companies (and they can be abominable), but fluoxetine (the generic name of Prozac) is amazingly better than the old meds, the tricyclics and MAO inhibitors that routinely caused horrible side effects and could be lethal in overdose. In contrast, treated with fluoxetine, decades of OCD symptoms will melt away, disabling panic attacks or severe depression fades, and patients who never responded to Elavil or Nardil find their lives transformed by 40 or 60 or 80 milligrams of this miracle drug. Bottom line, we can help so many people here, it's amazing.

It's not just the new drugs, either. We knew we would be applying the principles of the *DSM-III*—psychiatry's new diagnostic system, replacing the outmoded *DSM-II* from 1968—here at MMC, but we had no idea how damned useful it would be, what an explosion of new opportunities it would create. We have a new "brief psychotherapy" project, in which we are videotaping patient sessions for therapy treatments (forty fifty-minute sessions)—cutting-edge psychiatry research, led by our director. We offer cognitive behavioral therapy, multi-family therapy, trauma therapy, Hispanic women's groups, a dozen more treatments. It's a far cry from the one-size-fits-all model of infinitely slow psychoanalytic therapy for everyone from the old days.

Instead, our thousands of patients, with dozens of diagnoses, crowding our waiting rooms, need quick and accurate evaluations with reliable diagnoses; they need rapid and effective treatments that will alleviate their main symptoms, produce results in a reasonably short time, and—unlike psychoanalysis—not require several-times-a-week treatment for many years. With the *DSM*—imperfect as it is—you can make a diagnosis quickly and choose treatments that are likely to work.

What are the basics of the *DSM* approach? Essentially, it's simple. You empathically interview the patient, asking about current symptoms and stresses, their life situation, a brief summary of their life history; you get their medical history and current medications, and so on. You do a brief mental status examination, assessing thought processes and looking for any self-injurious thoughts and behaviors. Then you make a diagnostic formulation and begin to make a treatment plan.

In an hour or so you can get sufficient information to make tentative diagnoses and begin to come up with a treatment plan. *DSM* diagnoses follow a standard format. Major core symptoms must have lasted a certain period of time and need to have a significant effect on life functioning or cause significant distress. And the symptoms must not be better explained by other diagnoses. For major depression, say, a person needs to have five or more out of nine symptoms, lasting at least two weeks, causing distress or impairment, and the symptoms can't be better attributable to, for example, a medical issue, or substance use, or bipolar disorder.

Once you've mastered the method, it's relatively easy to apply this to nearly any patient. And for the main *DSM* diagnoses, it is highly "reliable"—in that two or more clinicians will generally agree that a patient has the same diagnosis. (How "valid" is the diagnosis? How much does one diagnosis relate to a specific abnormality in part of the brain or in the way the brain functions? That's another question entirely, and one that is far more complicated than in other areas of medicine.) But for our work at MMC, being able to make reliable diagnoses is a *huge* step forward.

A friend and I have just applied for a grant from the National Institute of Mental Health to treat patients doubly afflicted with schizophrenia and addiction. We're planning to start depression studies too, combining medicine and evidence-based psychotherapy. In just a few years, I've gone from being a clinic doctor to becoming the assistant director of the clinic. And just a few months ago, I became clinic director (just in time for the strike!).

Not to mention what's happening at home. As of summer 1989, my wife, Lisa, and I have a two-and-a-half-year-old daughter and a second child on the way. In her last trimester of pregnancy, Lisa has become enormous. She blames it on the meatball heroes she has been ordering for lunch in her job at NYU Medical Center, where she works as director of reimbursement. Or maybe it is the ungodly heat, which makes her retain fluids. Whatever the cause, she can barely drag herself out of the apartment, not to mention up and down the subway steps on her way to work at NYU.

And I worry: what if she goes into labor during the strike?

INVISIBLE PEOPLE

All spring the strike has been looming. First, we heard rumors from the people at the front desk, the clerks and secretaries, and the ladies in the record room. Then the social workers started talking. The maintenance men grew grouchy and defiant. Garbage sat in stairwells, and dust grew thick on window ledges. Nurses whispered about whether they'd work when the strike came.

"Contract's coming up," we'd be informed. "We may walk, we may not."

"Do you *want* to go out?" we'd ask. You couldn't get a straight answer.

The previous strike action was in July 1984, only three days after I started work at MMC. It lasted forty-seven days, and when it was over, the workers got nothing more than the hospitals had initially offered. Plus, they lost forty-seven days of pay. Five years later, we're told, some folks are still in debt.

"We got shafted," a maintenance man tells me. "The leaders, they led us on for nothing, we was used."

But our clerk Damon confides: "It's different now. We have a new leader, Dennis Rivera, he's a straight shooter, he knows politics."

"Plus," adds Janis from Medical Records, "the Catholic hospitals, they already gave in, they already gave 8 percent. All we want is what they got."

"That's right," says Damon. "If the Catholics can afford it, why can't the league?"

The League of Voluntary Hospitals, that is, which employs over fifty thousand union members. And which includes MMC. Just a week ago it was decided by the union: three one-day strikes; if those don't work, then they'll walk out longer. But it's complicated, because here (unlike most of the other hospitals) our nurses are union too, but in a different organization, the New York Professional Nurses Union, having split off from 1199 a few years ago, and because of a nurses' strike last winter at another hospital, they're already sure of a big increase.

Then, of course, it's an election year. An election for mayor in a year in which the city is trembling in one of its periodic convulsions, threatening to sink. Worsened by our multiple epidemics: AIDS, drugs, homelessness, crime. With no good treatments and over three thousand AIDS cases per year diagnosed in the city, plus thousands more not yet diagnosed, and one out of sixty women giving birth being HIV positive, the epidemic is out of control. The economy pitches and heaves, the Dow rises up one day and crashes the next. Streets fill with the homeless, and the rich can't find buyers for their million-buck lofts. Street violence spreads, and the cameras catch everything. In front of every subway station stands a candidate, squeezing the voters.

Not only is the city faltering; its hospitals are close to bankrupt. A year ago, the great Presbyterian Medical Center lost fifty million dollars, and even though it fired hundreds of workers it is still in danger of being taken over by the state. This year, they say, New York hospitals will lose at least $300 million. All over town, hospitals are cutting back programs and staff, closing clinics, turning patients away.

In the midst of all this, the hospital workers want more. The men who empty trash cans at night, aides who change linens and empty bedpans, clerks who type insurance information into computers, nurses in ICUs: They want more than they're getting.

Every hospital has an invisible army of workers. Wheeling patients; delivering medications and trays of food; keeping recalcitrant elevators running; down in sub-basements, fixing the boilers; up above, sealing the leaky roofs. Everywhere, they are cleaning. Endlessly cleaning, so much so that even a slight slowdown manifests immediately in dirty floors, discarded surgical masks and plastic debris, dark footprints, and long streaks of grease. It doesn't take much of a slowdown before you feel the city's grime overtaking the hospital, so that instead of a clean-scrubbed and shining refuge, somehow holy and cathedral-like, instead the IRT is crashing through the hospital's front door and winds are blowing garbage in from the Port Authority bus terminal.

Invisible people become visible the moment they stop—or threaten to stop—work. It's not that they don't deserve more. It's just that—if you believe the hospital finance people—there's no money left. Finance has always pleaded poverty. But now perhaps it's true.

THE FLOATING HOSPITAL

After the first day, there is a sense of euphoria. The workers return, proud to have been on TV, the doctors marvel over how well things ran without them, and administrators chuckle about how many of the nurses—50, 60, 70 percent—crossed the lines.

A few weeks later comes a two-day walkout, the second half of which is called off at the last minute. Nearly every day, in anticipation of the strike, we doctors have been meeting. Instead of our usual July discussions, orienting the new interns and residents, we focus on triage: Which services can be cut? Which staff is essential? Which doctors can continue to be docs, and which need to clean bedpans? Who takes the day shift, and who gets graveyard?

Oddly enough, this one hardly hurts. Sure, some nerves are frazzled; the residents in particular feel irked that they need to stay up all night and then work all the next day or when they are sent to surgery floors and are pressed into duty as aides or clerks while surgery residents stand around idle. The most annoying thing so far is that summer vacations are put on hold. Not to mention that even when the workers *are* back at work in the hospital, everything is slowed down.

The heat and humidity rise. Lisa can now barely walk. Now massively pregnant, she has stopped going to work. She rarely even leaves the apartment. Our building's elevator stops one floor below our apartment's entrance, and she has great difficulty dragging herself, plus the stroller, bags, and our squirmy two-year-old daughter, up and down the steep metal stairs. One day she calls me in a panic: she has taken Sarah a few blocks away for a haircut and can't make it back. Her hips are giving way. I rush out of the apartment and find a cab to bring us the three blocks home.

She's due in less than a month.

In early August 1989, the stakes rise. A three-day strike is called. My clinic is closed except for emergencies, and my staff is scattered throughout the hospital. A command center is set up in Administration to better allocate staff. This is almost becoming second nature. But our patients are suffering, running out of medicine. Already, at least a dozen schizophrenic patients have ended up back in hospital, and patients with PTSD, depression, and bipolar disorder are showing up in the emergency room suicidal, hallucinating, needing crisis services. All of the residents' educational programs are on hold.

Now, morale begins to fall, especially since we hear that the union and the hospital association aren't even talking. From uptown, we hear of serious sabotage, of toilets smashed and fires set.

When we hear that this outage will be followed by an open-ended strike starting in October and that we will then be assigned to twelve-hour shifts six days a week, tempers flare. It doesn't help that because of the financial crisis our annual raises were put on hold and our health benefits cut. Or that new hiring is frozen until further notice: if anyone quits, we can't replace them, no matter how essential their job might be.

UP ON THE ROOF

Lisa and I live in a penthouse on the Upper East Side. Or so we say. Actually it's the old maids' rooms from the days when the top floor was sooty and hot, a miserable place to live, in a creaky prewar building, without a doorman or the other amenities of the posh buildings over on Park Avenue. The forgotten maids lived in tiny rooms, which we now inhabit. The greenhouse roof often leaks, water splattering our furniture, and the wide wraparound terrace floods during downpours. But it's a magical place.

We bought it at a bargain price just after I finished residency, after the previous owner died of AIDS. Taking possession, we found a green-shag-carpeted wreck with mirrored walls inside a flimsy structure of cinder blocks and corrugated tin, surrounded by a buckling tar roof. We did some essential repairs and bumped out the kitchen and second bedroom. We installed tiles over the tarpaper on our terrace and then bought planters, which we filled with birch trees and soft white pines, and put in a recirculating fountain and a plastic slide and sandbox for our two-year-old daughter.

It's a slum, really, a rooftop shack with a 13 percent mortgage rate, but it's home.

One afternoon—in the midst of the three-day strike—I hear a roar and look out to see a vast tide of people coming through the park toward the hospital. It's mid-August, two weeks before our baby is due. At the front of the approaching crowd of people in blue scrubs and white coats and brown maintenance uniforms is a certain brightness, a seeming focus and clarity to the atmosphere.

An amplified voice crackles over the roar of traffic, a familiar voice of booming imprecations, and I can make out, a hundred yards away by the dry fountain, the figure of Jesse Jackson surrounded by union leaders.

After Jackson finishes, the crowd again surges toward the hospital, toward the dozens of cops and baby-blue-painted police barricades, a human tide that washes past the emergency room and the cafeteria and the loading docks, chanting and beating drums and blowing horns, until the entire Manhattan Medical Center seems to tremble and roll, ready to rise from its foundations and float into the sky.

It's dark by the time I get home. Drenched in sweat, I take the elevator to the top floor, then climb paint-encrusted blue metal steps, balancing a soggy box of pizza, and go inside. The rooms are stifling hot, despite the air conditioner's rattle.

I find them outside on the terrace. Sarah is in the sandbox, dumping sand on a plastic dinosaur. Lisa is sitting on one of the splintery Adirondack chairs. I lay the pizza box on the wooden outdoor table and separate the floppy cheesy slices.

I bring out plates and forks and napkins and tall glasses of decaf iced tea, and we eat, talking about the strike.

Finally, it starts to cool down, to turn into a beautiful night. We see neighbors around us, the lights coming on in their apartments, the sky overhead. The moon often rises over the rooftops, just above the water tanks to the east, but so far not tonight—just dark sky and a few high cirrus clouds.

Eventually our daughter is sleeping in her tiny room, and I am sitting outside with my worn-out pregnant wife, drinking decaf iced tea from sweaty glasses. We are talking about all this stuff that is happening. It's damned confusing. On the one hand, perhaps as a consequence of the success of the *DSM*, the ways our new psychiatric diagnostic manual has revolutionized psychiatric treatment, we psychiatrists have finally been embraced into the healthcare mix, along with medicine, OB/GYN, surgery, pediatrics. We're fighting alongside them for a better outcome for our patients—and sometimes fighting *with* them for a bigger piece of the pie. And Lisa, working at NYU and Bellevue, faces her own set of intractable problems. Too many hospital beds, too many uninsured people, crazy insurance rules, chaotic community care.

It's a fundamentally broken system. Compared to just about every European country or closer by, Canada, costs in the United States are twice as high for far worse outcomes. Not that a national health system would be perfect either, but we'd be able to provide more care to more people without so much being siphoned off to insurance companies.

And in a way, both of us are part of the problem.

The over-bedded hospitals—like NYU Medical Center—admit too many patients and keep them too long, because it pays. In clinics like mine at MMC, the more patients we see, and the more frequently we see them, the more money we make. Medicine, surgery, psychiatry, hospitals, clinics, whatever—we all can game the system. The fact is, we *have* to game it; it's the only way to keep going. Ideally, we'd distribute the care better and give to more people: there are so many people who can't access health care. And so much wasted money!

Plus, if you look at any one part of the system, it's even crazier. If they—meaning the MMC hospital staff, the union folks—win, we—the doctors, that is—lose. They get their raises; our salaries get cut. And yet, even with the craziness of the strike, even with the financial crisis, everything still seems possible. That's what's so weird, our strange irrational hopefulness.

Then our two-year-old daughter comes out through the living room's sliding door and stands barefooted, wearing her Little Mermaid nightgown, clutching her bedraggled red Elmo. Past the fountain, the plastic slide, the green plastic sandbox in the shape of a turtle, she stands in front of us.

"What's up, Sarah?"

"I couldn't sleep, you were talking."

"Sorry, honey," says Lisa. She doesn't move. "Dave, get up!"

In my arms, Sarah's little body is sweaty, her Elmo, a furry red creature, has a pungent child smell. I carry her to the parapet at the edge of the terrace, and we look upward, scanning the heavens.

"Where's the moon?"

Sarah looks over the water tanks, and the dark roof lines. "Moon sleeping!" she says.

"Yeah, I guess it is. Hey kid, time for bed."

"No," she says. "No way!"

Lisa laughs, holding her belly.

I sink onto the Adirondack chair with Sarah on my lap, our squirmy little curly-headed girl, and hold her as she slowly falls asleep. Then I take her inside to her room underneath the building's wooden-slatted water tank, with an empty crib already waiting in the corner.

THE UNCERTAINTY PRINCIPLE

"Let's be honest," one of my colleagues says. "Let's not bullshit each other. Where do we stand? Are we management? Or are we labor?"

August 20, 1989. An ad-hoc meeting the last day of the three-day strike. Ten days until Lisa is due. It's getting personal. What if this all comes down when Lisa's in the Labor and Delivery Department?

"As doctors, our first loyalty is to our patients."

"Of course, of course," everyone echoes.

"But are we being *used* for that? Strictly speaking, we're scabs. We're being used to crack the union. If we don't work, we're fired."

"We're *professionals!*"

"So what? That means we just get screwed. Maybe we should join a union ourselves."

"Why the hell not!"

We all know, of course, that none of us would.

Here we are, physicians on the Lower East Side, working with Medicaid and Medicare patients, when we could be making a lot more money somewhere else. Are we not, most of us, grandsons and granddaughters of socialists from Russia and Poland, of railroad workers from Ireland, of Italian stonemasons? Didn't our seamstress grandmas join the International Ladies Garment Workers Union? Didn't our factory-worker granddads risk cracked heads on picket lines? Deep down, aren't our sympathies at least as much with the strikers as with the bosses?

A few days later, we are at our unit directors' meeting, planning for the open-ended strike starting in a month.

"These damn hospitals, they're no different these days than IBM or AT&T. Same damn money-grubbing CEOs! We're just pawns, we're completely expendable!"

"Face it, if we don't become businessmen like them, we're always going to get screwed."

"Then we've lost it all. It's the fact that we *aren't* part of management, we *aren't* businessmen, that gives us any goddamn integrity, that's what makes us better than both sides."

"Why aren't they even talking?" one of the younger attendings asks plaintively.

Now that we are facing the open-ended strike, the conversation is getting weird: the talk of factory workers, not professionals.

An older doc snorts. "It's all theatre," he says. "The Hospital Association and the Union are in this together." He's been around. They're working out how to get the best press possible now that the election's coming up, he explains. They've already figured out exactly what settlement they're going to make. "They're just dragging things out so the voters will put pressure on the politicians who are up for reelection, to cough up more money for the hospitals. It's a shell game!"

Then we get down to business, making schedules for the infinitely more complex open-ended strike, trying to avert disaster. Funny thing is, despite all our complaining, none of us would dream of leaving.

EIGHT PERCENT

Luckily, Lisa goes into labor between strike actions. When her waters break on the morning of August 29, we cab the eight blocks to Lenox Hill Hospital. The floor is frenetic, as though the entire city is giving birth at once. All the rooms are full, so Lisa is consigned to labor in a utility closet, surrounded by mops, brooms, cleaning supplies. Outside a Hasidic woman screams: "*HaShem! HaShem!*" The forbidden name of God. And curses her husband in Yiddish.

But all the nurses are here, so we don't care.

Only fifteen minutes later, Lisa's OB/GYN shows up. Soon we have an enormous meatball-hero-fed baby boy, weighing in at ten-and-a-half pounds.

The open-ended strike is called for October 4, when he is a little more than a month old.

I struggle to get out of the apartment that morning, kissing Lisa and our newborn son, Ben, goodbye, dropping Sarah off at her preschool down Lexington Avenue, then rush downtown.

It's not until I walk into the clinic that I know for sure. The entire front-desk staff is back at work, beaming, laughing, bragging, euphoric. Turns out they came to terms at 1 a.m. last night, hours before the 5 a.m. deadline for the open-ended walkout.

"We got our eight!" they say. "We won! Got everything we asked for!"

It's a bright early fall day, the sky a brilliant and thoughtless blue. At lunchtime, the maintenance men stand around outside by the cast-iron gates to the park, flirting with the clerks. For the first time in months, they're grinning.

Things gradually go back to normal, more or less. Within a week or so the hospital looks as gleaming and clean as ever. The residents need to be supervised, and since we are understaffed by three doctors, I have no one to teach them. Also, because of the hiring freeze, they have to carry enormous caseloads, and I spend hours trying to figure out how to assign new patients who are desperate for treatment.

But it's fine. Lisa is fine, our huge son, Ben, is fine. He has chick-down hair that stands straight up, chipmunk cheeks, and permanent smile. Sarah is adjusting in her own way, putting slices of bread in the VCR, methodically dumping sand on the rug, and grudgingly, gradually, learning to love her brother.

One week, five therapists in the evening clinic hand in their resignations; they can get better money elsewhere. Our best clerk at the front desk, the only one who can keep up on the computer, quits ten days after the strike is settled: she's moving down South. Even 8 percent can't keep her interested.

Thom Singletary and I and one of our inpatient psychiatrist friends, Madden, the son of an Irish cabbie from Gerritsen Beach, Queens, go for lunch at the new noodle place near Tompkins Square Park. We slurp our guilin rice noodles and chengdu dan dan noodles and talk about how, after several years of growth, it seems that we are entering a new phase. The crisis in our clinic, on Madden's inpatient unit, in the hospital itself, and here in New York City, maybe it is just a local expression of things that are starting to shake up the entire American health care system. A system that always thought it was working pretty well, that it didn't need more than tinkering.

Finally, as the holidays approach, peace returns. We get the OK to fill some of our empty lines; we place ads and begin to interview. Good applicants, too. Despite the economic gloom, a lot of people want to work here, now that the Lower East Side of Manhattan is getting gentrified; suddenly this is a fashionable place to work, to go clubbing, to live. The residency application season begins, and a new crop of senior medical students, fresh-faced and enthusiastic—better and brighter than ever—appears at our office doors. A pleasant amnesia quickly overtakes us.

And we start going back to our previous optimism and excitement: things will work out. The health care system is certainly in urgent need of reform, but it must be possible to figure a way out of this mess, for my clinic, for my department, for MMC, for the country at large. Anyway, we're doctors and therapists; we can't fix the system—we just have to treat our patients to the best degree possible with the resources we have.

We get back to supervising residents in our short-term therapy project, to submitting research presentations and grants, to learning about all the new psychiatric drugs that are coming out—new SSRIs in particular, highly selective drugs that work on so many disorders, everything from depression to panic disorder to bulimia to compulsive gambling . . . we've got Prozac, sure, but now they're talking about Zoloft and Paxil and Luvox, too! We're back to the excitement of working in an era of unprecedented progress in psychiatry.

Now able to walk again, Lisa starts working part-time at NYU Medical Center, going in three days a week. Every morning we pray that our babysitter will show up, and we trade off dropping Sarah at All Souls preschool. Sarah tells us she loves being a big sister. Despite hitting Ben with a pot lid, she causes no discernable damage. In short, all is well.

But then, in the first week of November 1989, I am startled to attention by item no. 4 on the executive committee agenda: "Strike Planning."

"What's this?"

"Didn't you hear? Nurses' strike."

"*What?*"

"Starting Thanksgiving Day."

All around the conference table, faces turn. Should the nurses go out, we know all too well who will be called in to cover.

"What are you talking about? They settled!"

"No, the nursing contract's not up until December."

"But they got their raise!"

"They want more."

"Oh God!" we all say. "Not again!"

Our director smiles ruefully. "We'll put off planning until next week. Should be second nature by now."

CHAPTER 6

Reinventing the Egg

Translating the *DSM* Across Cultures and Languages, 1990–1994

Today, it's Dr. Jenna Li, newly arrived from Guangdong Province in southern China. Pallid and nearly expressionless, in a grimy white coat, too-short gray pants and scuffed loafers, she trots out a simple story of a man who won't take his meds.

I've heard his story before. Two years back, Dr. Ranu, an Indian-born resident, presented him as a psychodynamic puzzle, a man awash in neurotic resistances, in desperate need of psychoanalysis. Last year it was Dr. Soo from Seoul, Korea, who saw him as embodying neurobiological connections gone awry, with reverberating brain circuits linking pain, fear, and depression, and who recommended three-times-a-week electroshock treatment.

The man waits outside our conference room. Mr. Illich, I'll call him, a fifty-something-year-old graduate of Moscow State University, once an engineer and now an unemployed cab driver, spends most of his days wandering through Manhattan Medical Center, from neurology clinic to cardiac echo lab to the diabetes center, occasionally dropping into the psychiatric services, where he has for the past several years served as a Rorschach test for each class of psychiatry residents. Or an in vivo Kurosawa movie, *Seven Samurai on the Lower East Side*, each psychiatric warrior perceiving a different drama.

Suffice it to say, whatever theory is applied to this man with unremitting depression and high levels of "suicidal ideation," and whatever treatments ensue, Mr. Illich remains, well, Mr. Illich. The residents try their best during medication visits and therapy sessions; they draw blood for obscure lab tests; they order MRI scans, EEGs, and neuropsychological tests; they meet with his wife and two teenage children; they write reams of prescriptions . . . and nothing. Actually, the residents do fine: they complete their year of outpatient psychiatric training and eventually graduate, getting good jobs or research fellowships or opening private practices.

And Mr. Illich remains Mr. Illich. Depressed, immobilized, in pain, yearning to die.

Until now, each year I have had a faint hope that the new resident's latest theory would bear fruit and that somehow Mr. Illich's intractable headaches and chest pains, his gastric reflux and neuropathic foot burning will fade away and that he will find a reason to live. But today, as Dr. Li presents Mr. Illich yet again at case conference, my optimism fades.

I look around the room at her fellow residents, who come from all over the world, and at the staff doctors, the social work therapists, and the psychologists, who come from across the United States.

Is it really possible to cross the divide? For a young doctor from one culture to connect with a patient from an entirely different part of the world? An internist or endocrinologist can give anybody from anywhere antibiotics or diabetes meds and be confident of the results. But a psychiatrist? We attempt to heal unseen wounds with empathy, bridging differences to make emotional connections. Sometimes the tsunami of stories and the complexity of multiple cultures here is amazing. At other times it's impossible.

Today seems the latter. Which is a shame, because I love working here at MMC—the clashing cultures, the poetry of the streets in a hundred languages, which we see every Thursday at our case conference.

It's a weird thing—our patients are stigmatized, ignored, othered, expelled from their families, banned from emergency rooms and clinics. We are here to help them. But *we*, their doctors and therapists, are stigmatized too. We are "shrinks," they tell us, witch doctors, alienists; they carve out our insurance reimbursement so our services are barely paid for. Yet here at the clinic, at MMC and elsewhere, we try to erase all that.

While by 2022 categories our staff is "white," at the time we see ourselves as Italians, Jews, or Irish, as red-diaper babies, doctor sons of Queens cab drivers, gay refugees from the Baptist South, children of Holocaust survivors. Our support staff is largely Caribbean American, and our patients come from everywhere. The dynamics are complex. On the whole, our patients are thrilled to get compassionate care, notwithstanding cultural differences. Particularly our patients with "comorbid" schizophrenia and substance use disorders who have been entirely "othered" not only by their families and friends but also by substance abuse and psychiatric treatment programs, for whom we set up an innovative dual diagnosis program. At MMC we work with hundreds of gay men during a time of great fear, many infected with HIV, and with innumerable unhoused patients who are despised and hounded.

We try to see things from our patients' points of view; we develop innovative group therapy approaches with peer counselors for minoritized patients with severe illness and disability. And we do research on incorporating patient preferences into hospital care. Our clinic houses a local mood disorders support program and a community-based peer support program for people diagnosed with bipolar and unipolar mood disorders. We work with innumerable ethnic, racial, social, and gender identities, many who have had profound experiences of stigma, discrimination, and marginalization and who rebuild their lives by creating community.

"Listen to your patients," I tell my residents and psych interns. "Connect to them, see them as people, try to understand their experiences. And if you do listen, you will hear amazing things. It's for diagnostic purposes, sure, but you also have to understand their vocabulary to connect empathically so you can provide psychotherapy and get them to take their medicines."

And I must admit, my interest goes beyond that—their very poetry carries me away.

"My father is Malcolm X, my mother's Tina Turner," we hear one Thursday morning at our weekly case conference. A rail-thin man, Alial exclaims, "I invented crack! It'll kill all the hookers and junkies on Fourteenth Street, I make it out of trees!"

Another Thursday, a Dominican elder confides, "The Virgin Mary, she anger for the spirit man coming to me. I wake worrying, I crying, will I have a spirit baby come?"

An Australian merchant sailor comes to our walk-in clinic. He is being assaulted by "forces of magnetism released by construction in the city." Plus, he believes that wings are sprouting from his feet.

"What do you make of this?" I ask. "What accounts for these experiences? What disorder may they have? What would be your treatment approach?"

But Dr. Li confounds me. So shy, so timid, so pale, so quick to agree with whatever an authority figure says, she is always scrawling something on a pad of lined paper. Yet after she's been working in my clinic for three months, I have no clue whether she understands what we discuss in case conference or in our weekly supervision meetings.

Now she finishes her presentation, rifling through her grimy notes. "My question, my question is . . ." She stumbles. "Can we . . . can the doctor . . . can . . . ?"

I raise a hand. "Let's interview the patient."

She goes to fetch him. A wide stooped man wearing cheap trousers and scuffed shoes without socks, Mr. Illich looks a good decade older than his actual fifty-three years. His watery eyes appraise us. He rubs the stubble around his goatee while my friend Thom Singletary, a Black psychiatrist who grew up in the South, the first member of his family to finish high school, let alone college or medical school, plies him with questions.

For the next twenty minutes, we hear about Mr. Illich's education at Moscow State University, his sterling academic record followed by struggles to find work because of antisemitism. His years of unemployment, near bankruptcy, the short-term jobs in Norway, Saudi Arabia, and Dubai, and the inevitable decision to leave and come to the United States in the early 1970s. Much suffering with life as "a Jewish." And here, in Brighton Beach, Brooklyn, his children struggle but turn American, his wife trains to become a pharmacist's assistant—only he is left behind, in agony.

This is the painful side of Thursdays: the encounter with irremediable suffering, with wrecked lives. It is awful seeing Mr. Illich talk before our group, our doctors and psychologists and social workers, to be in the presence of someone so vitally wounded. I can see that Thom holds back, not wanting to delve for too many specifics. But it's interesting to hear his questions. Thom is tall, thin, very dark skinned, and has a courtly, shy manner. He lives up in Harlem and never tells us what he does on weekends. He and I and Madden, one of three psychiatrists on the inpatient service, often go out for lunch at a local Polish restaurant, where we sit out back in their garden, eating pierogis and kielbasa-and-egg breakfast specials and sipping iced coffee and laughing about the latest hospital craziness. Thom is more than a bit of an enigma, and he seems to view us with equal measures of curiosity and amusement. He is kind, as ever, to Mr. Illich.

Then, during the discussion period, I ask, "Mr. Illich, is it true that you don't take your medicines?"

He looks at me in desperate abandon. "No, I take, I take."

Dr. Li raises a hand. "But your blood level, Mr. Illich . . ."

"You got the blood level of his amitriptyline?" asks Thom.

"Yes."

"And?"

"It was zero."

Illich slumps even more, defeated, utterly humiliated before us.

Dr. Li escorts him out and takes her seat again. We discuss the stresses of immigration, the presentations of somatization disorder, and whether imipramine or nortriptyline would be a better medicine choice than sedating amitriptyline. But what difference does it make if he won't, for some mysterious reason, take a single pill?

"So, it is simple," says Dr. Li, suddenly smiling. "We will tell him to take his amitriptyline; he will get better. Thank you, Dr. David and Dr. Thom."

If only it was so simple. I am left speechless by her final comment: does she understand anything about her patient?

The Manhattan Medical Center, where I have been working since graduating from residency training eight years ago, clearly occupies a different universe from the late, lamented Payne Whitney. The Payne Whitney Clinic was in the city but not of it. It was always remote and cerebral, proudly elitist, whereas MMC is intense and practical and canny, embracing (or overwhelmed by) the wildness of the streets.

Shortly after starting at the MMC's psychiatric clinic as a junior attending physician in July 1985, I came to realize that every urban crisis, every epidemic, and every disease and social problem finds an echo in the hallways of Manhattan Medical Center. So does every problem of the world. Our residents come from all over the globe, as do our patients. We are awash in stories here, told in a pure poetry of suffering and survival. Broken stories, you might say. Whatever school of psychiatry we follow, whether we embrace cognitive therapy or psychoanalysis or psychopharmacology, it is these fractured tales we hear, these broken stories we try to repair, by translating them into the simple terms of the *DSM-III*. Every day here we are exposed to intense suffering and unlikely survival, every day here we are awed, overwhelmed, aggravated, inspired, and whipsawed by infinite contradictions. We labor continually to patch together shards of broken lives.

The Iraqi woman who has been forced into prostitution, so crushed that a fountain of tears literally flows down her cheeks. She clearly has severe major depression, and we are convinced that she can be helped with our new *DSM-III*-inspired treatments, by the brand-new SSRI medications, by "evidence-based" trauma therapy—and that a new story will gradually emerge, a new life imaginable. The East Harlem incest victim, the Japanese jet-crash survivor, who both have severe post-traumatic stress disorder, the gay refugee from a fundamentalist Pakistani village or a town in Oklahoma, who has adjustment disorder with mixed emotional features—one moment, we are certain we can help. The next, we know it is going to be a huge struggle.

At our best at MMC we inhabit a miniature United Nations. Translation is our default mode. Often this process goes three ways, between the world of the immigrant patient and the world of the immigrant resident—the doctor-in-training—and the world of the American-born attending physician.

And at best, the *DSM* is our common language. For all its imperfections, it allows our doctors and therapists to translate incomprehensible suffering into comprehensible disorders and to choose rationally among possible treatments. It provides a shared vocabulary for suffering and potentially a path toward cure. At best, the *DSM* is therefore profoundly antistigma, antiprejudice, and antiracist, at least as we try to apply it at MMC. By quantifying and defining suffering, we can make plans in partnership with our patients, to work together toward common goals.

At its worst, though, it's chaos. At least in part because of the residents' enormous workloads, fifty or more psychopharm patients, dozens more in their group and individual caseloads. Continuity of care is our gold standard, yet it may be impossible to maintain. Literally months may pass before coming back to a case.

But beyond that—we come face to face with the immense challenge of communicating across cultural and language divides. Do the stories—or the sense of the stories—get lost in translation? Which is what I experience with Dr. Li.

Over the coming months, the end of 1990 and beginning of 1991, I meet regularly with Dr. Jenna Li to supervise her many cases. I do learn a bit more about her: how her parents were caught up in the Cultural Revolution and exiled to the countryside for reeducation, how she didn't see them for eight years and was raised by her grandmother. I learn how after medical school she decided to emigrate to the United States, how she came on a tourist visa and ended up volunteering at the medical laboratory of a Chinese doctor in Texas. How she studied for Step 1 Exams, receiving an excellent score (she is clearly very smart!), enabling her to apply to residency training in psychiatry. And how she arrived here at MMC and is shocked by the depravity of the city. Now she struggles with her American visa and finances and one housing problem after another. Finally, she ends up living in a residence for single Catholic women near Penn Station, where she shares meals with soon-to-be nuns. She sends most of her paycheck back home to Guangzhou.

That is all fine, and the usual for our young doctors—and amazing. I could never make the reverse move. Over the fall and winter, I gradually see her emerge from her shock and withdrawal. A quiet, thoughtful person with a poetic sensibility, she also seems to have a sly sense of humor, though often I can't tell if she is being serious or deadpan joking.

It's not simple working with Dr. Li, though. Some of it is vocabulary, comprehension, but there's more to it than that. Sometimes I literally can't understand what she says. I comprehend her words but not her sentences, or if I can understand the sentences, I don't understand the meaning. She just thinks so differently from those of us born and educated in the United States. Maybe it is something about Dr. Li's schooling, from what she tells me was a rote-based Chinese curriculum. Or perhaps it stems from the structure of the Cantonese language. Or maybe, I wonder, it reflects quirks of her personality.

An example: one morning I hear from one of her patients who has run out of lithium and other medications. A middle-aged Nigerian lady, diagnosed bipolar, Mrs. Ayodele has called the clerk, who forwards her through to me. I try reaching Dr. Li, who doesn't pick up in her office and doesn't answer several pages.

After lunch, I find Dr. Li by the Xerox machine, calmly copying some journal articles.

"Did you write Mrs. Ayodele's prescriptions?"

"Yes, I write."

"Have you written them already?"

"No, later I will do it."

"She's coming to pick them up, so you should write them before you leave today."

That evening, as I'm preparing to leave, I ask: "Did you give Mrs. Ayodele the prescriptions?"

"Did you see me give her the prescriptions?"

"No, I didn't see you give them to her, I was in my office all afternoon."

I eventually understand that Jenna means she *didn't* give the prescriptions to Mrs. Ayodele yet. In fact, she hasn't written them yet. Why doesn't she just say so?

Meanwhile, Mrs. Ayodele patiently sits hour after hour in the empty waiting room. Finally, I stand over Dr. Li while she writes out the scripts and watch her hand them to the patient. She does not seem to think any apology is needed.

That is a small example, I guess. More troubling is that Dr. Li doesn't seem to understand the *idea* of psychotherapy. Her supervisors complain that her sessions are short, only ten or fifteen minutes, and her chart notes only have a line or two of text. When asked, she says she doesn't understand what to talk about for forty-five minutes in a room with a patient. The same could be said for her supervision sessions. There are a lot of patients, true, but there's not much to talk about regarding her work with any one of them. So, to fill the time, we end up telling each other stories.

She tells me more about her family's history before the Cultural Revolution, how her grandfather, a violinist, lived in Paris for a decade, how her grandmother was a professor of German literature. I find myself telling her about my own family. How my mother's ancestors came to the United States before the Civil War. And how, on my Dad's side, immigrant memories were still raw. I tell her how Dad's parents were penniless refugees, living briefly here on the Lower East Side just a few blocks from MMC. And how my father's father began working as a peddler in the Ohio Valley.

I tell Dr. Li how my grandfather spent years traveling through coal-mining towns in the folds of the Allegheny Mountains, until at some point he was taken in by an older single man, Finkelstein, a childless older Jewish man who ran a dry-goods store in the town of Dillonvale, Ohio. "Eventually they opened up a store together: Finkelstein and Hellerstein, Outfitters from Head to Foot."

Dr. Li's face lights up as I tell how Samuel married Celia, his childhood sweetheart from the same city in Belarus. And how they settled down and raised five children, before moving north to Cleveland on the cusp of the Great Depression. It seems to strike a chord: the theme of family resonating across the world. She seems to relax a bit, to open up. She tells me about her parents, her mother's work as a physicist, her father as a mathematician, and how they both swept streets after their Maoist reeducation. How hard it was when they were away for so many years.

"Thank you, Dr. David," she says, at the end of our supervisory hour. She picks up her notepad and walks out down the hall, leaving the door open behind her.

We all have our reasons for ending up at MMC, here on the Lower East Side of New York City. Our patients come from Peru and Ukraine and Cambodia and India and every part of the United States. Our doctors-in-training likewise hail from around the country and around the world—from Romania, Nigeria, Ireland, Croatia, Israel, and Russia, to name a few homelands. In a way, most of us are refugees. Dr. Liskaya, a thickset Russian woman in her fifties, with a PhD in physiology and an MD, is doing her second residency. She was caught up in the gulag, but no one has courage enough to press her for details. Dr. Patel, Indian-born, is brilliant and apparently comes from money: the department secretary complains that he never picks up his paycheck. Dr. Mattias O'Linsky, a Scottish Jew—I love the apostrophe, it seems so apt—somehow finds housing in a residence for Scandinavian merchant sailors and is obsessed by the idea of measuring the effects of altitude on the mood of an expedition climbing Mt. Everest.

Despite our differences, we have to make it work somehow. Sometimes we seek common ground in the hospital, and the byproduct is a sort of cultural glasnost.

At our Thursday case conference, one of our brightest residents, a Croatian, finishes a presentation of a twenty-seven-year-old Irishman with severe OCD in which he has integrated the patient's clinical history with the latest thinking in neuroimaging, brain receptor chemistry, and cognitive neuroscience. "It is possible," he says, "that here we are reinventing the egg!"

Everyone laughs, and the resident blushes as his colleagues trot out other proverbs: "Don't change eggs in midstream!" "Don't cry over spilled eggs!" and so on. A mixed metaphor but brimming with truth, explaining our manner of practicing medicine across cultures and how we see ourselves anew.

From Dr. Patel, we learn that in India and Pakistan people often must wait years to see a doctor and that few medications are available (in psychiatry, for instance, often just the old-fashioned 1950s drugs chlorpromazine and amitriptyline).

Dr. Gutierrez, born in Mexico City, explains that when Mr. Ramos complains that he has "bad blood," he believes that anger actually curdles his corpuscles, attracting biting insects. Dr. Hernandes, from the Dominican Republic, disputes one patient's diagnosis of "panic disorder"—instead, he insists that she suffers from *ataques de nervios*, a Latin American condition brought on by stress and anxiety. And Dr. Soo presents a man, recently immigrated from Seoul, with *hwabyung*, a Korean disorder in which insomnia and panic arise from suppressed anger.

The paradox is this: while every doctor yearns to achieve an "international reputation," no hospital wants more than a certain number of international medical graduates. International standing as a practitioner or researcher implies a sterling reputation, but the prejudice of the American medical establishment is that good training programs should be able to fill their slots with U.S. citizens. A patient, on hearing his doctor's accent over the phone, refuses to be treated by him and demands "an American."

"Oh, I hear you have mostly foreign grads," sneers a colleague from a fancy uptown hospital. "Are they any good?"

Somehow, I realize, the multicultural swirl of MMC is familiar to me, even reassuring. It reminds me of childhood in Ohio.

Not my suburban neighborhood, though: that was a place of danger. The Catholic kids who lived behind us drew swastikas in wet cement between our garages and wrote "Kill the Kikes!" The boys next door, whose father ran a funeral home, ambushed us with snowballs packed with rocks. The very priest from St. Ann's Church School, a quarter-mile away, refused to accept Pope Paul IV's 1965 encyclical *Nostra Aetate*, which had absolved the Jews for the death of Christ. So, the Catholic boys taunted us:

> Roses are red,
> Violets are bluish,
> Jesus is dead,
> I'm glad I'm not Jewish.

The Protestant kids never attacked or insulted us but disappeared on weekends to their country clubs, where Jews never set foot. And downtown, at the

Cleveland Arena, after a hockey game, we were chased by a mob of Black kids and hassled after stopping at a Carnegie Avenue Burger King.

Everywhere, even on our own street, we were reminded of our uneasy status as outsiders, immigrants, invaders.

Instead, MMC reminds me of the world of University Hospitals, in downtown Cleveland, where we spent a good part of our childhood. The hospital was full of refugees, every culture, race, ethnicity, both patients and hospital staff. Our father, Herman Hellerstein, was a cardiologist, a pioneer in the rehabilitation of heart attack survivors, and often on weekends, to give our mother a break, he would take one or more of us (there were six in all) with him to the hospital. We would develop film in his darkroom, or use treadmills and bicycle ergometers in his research lab, or put on musty white coats and follow him on rounds to the Coronary Care Unit. And the hospital followed him home: eager to learn from him, doctors from all over the world came to visit. In 1950, a Dr. Toyomi Sano came over from Japan—so recently our country's bitter enemy—to study cardiology. He stayed for three years, and he and Dad, a newly discharged U.S. Army doctor, became close friends. Together they wrote a book on congenital heart disease, and the Japanese edition was given a prominent place in Dad's study.

During the tense post-*Sputnik* years, a Russian, Dr. Susoyev, came to visit, as well as doctors from Germany and Nigeria and Argentina and Egypt. Dr. Hahter-Khan from Hyderabad, India, brought gifts of papier-mâché camels and elephants and riddled us about how to ferry a tiger, a wolf, and a lamb across the Ganges River without any of the animals being eaten. There was even Dr. Montarez, a whippet-thin tight-lipped doctor from Chile, during the regime of Augusto Pinochet, who attempted to justify the murders of Salvador Allende's supporters.

This cosmopolitan parade toured the hospital and lab, meeting the research faculty, many of them World War II refugees from Germany, Austria, and Poland. Then we watched them across the dinner table, hearing their strange accents, answering their questions, as we all dined on Mom's meatloaf or chicken à la king. Maybe it was an illusion, but it seemed that whatever political or religious differences our countries or peoples might have, they were nullified by the common pursuit of healing.

Thus the origin of my fantasy of medicine as a meeting ground, transcending tribe, hatred, prejudice. Not always, but at its best, which perhaps is what has brought me here. Not that the MMC is entirely conflict-free. At times, my own clinic echoes with ethnic strife. Our Croatian resident refuses to be supervised by a Serbian doctor. One day I hear yelling and come out of my office to find a Russian resident and a Ukrainian psychologist who share an office almost coming to blows. Unable to broker a truce, Franklin, our administrator, has to move them to separate offices. And language differences can certainly fray nerves. The Cambodian refugee's broken English baffles her Hungarian doctor. The young resident from Barcelona, fluent in Catalan, does not speak the same Spanish as her patients from the ghetto of East Harlem. Doctors from the Communist bloc are often amazed by patients' willingness to reveal secrets to their doctors.

Eventually, I give up trying to convince Dr. Li to do psychotherapy. The concept just doesn't seem to translate. Instead, I say, "You like stories, Dr. Li. Just get them to tell stories. Mr. Illich, ask him more about himself. Ask him about why he became an engineer. How he fell in love with his wife. You can help him—" I am about to say, "change how he sees himself." But I lack confidence in that phrase. Instead, I say, "tell his story."

Dr. Li says nothing. As always, she gathers her papers and walks out of my office, leaving the door open behind her.

In the coming weeks, her caseload explodes with one crisis after another. Mrs. Wojciechowski, from Poland—major depression, recurrent, severe—is planning to commit suicide and has been accumulating pills. Two of her patients have terminal AIDS—a gay man and a woman who mainlines heroin. Both have AIDS-related dementia, from infection of the brain by the human immunodeficiency virus. A single-room-occupancy hotel on the Bowery closes precipitously, leaving several of Dr. Li's schizophrenic patients homeless for weeks. Dr. Li is overwhelmed. There is no time for her stories.

In late spring, Dr. Li makes plans to turn most of her cases over to an incoming third-year resident. I finally get around to asking her about Mr. Illich.

"He's much better," she says.

"Impossible!"

"Not impossible. He is very good. He is not so depressed, major depressive episode in remission, and now he is starting to take a computer course."

Dr. Li tells me that she has changed his medicines. Without asking me, she decided to put him on Prozac, and when that showed limited effectiveness, she decided to supplement it with lithium. Fair enough, I guess, for a patient who has treatment-resistant depression. In *DSM* terms, Mr. Illich's diagnosis could be further refined: recurrent major depression, severe and chronic, with melancholic but without psychotic features, and without full interepisode recovery. In such cases, there is evidence for benefit of different strategies. Maximizing drug dose, switching to a different medicine, supplement with a second medicine like thyroid hormone, lithium, etc. Each of these has been shown to help some people. Except for one thing: Mr. Illich's so-called noncompliance.

"You don't actually know if a person has treatment-resistant depression if they never take their medicine."

"No, he takes, he takes it now." She hands me Mr. Illich's chart. The notes are still only a few lines, her faint handwriting almost impossible to decipher. She turns a page to show printouts—blood levels for both of his medicines. There is no real reason to get blood levels of Prozac, but there they are before me: fluoxetine and its metabolite norfluoxetine, 81 and 133 nanograms per milliliter respectively, drawn on two different occasions. And his lithium level is 0.4 milliequivalents per liter—on the low side, but respectable.

Mr. Illich is taking his medicines. I am utterly amazed. "What changed?" I ask.

"What you tell me to do," she says. "I ask him to tell his story." She hasn't understood much of what he had to say about his oil refinery work in the Middle East, but she understands how hard he found it to be separated from his family for so long. She tells me that she told him about herself too, how she was separated from her parents when she was a child, during the Cultural Revolution. "For eight years." She says, "When I tell him that, he begins to cry."

When residents go into their fourth and final year of residency, we encourage them to keep some of their most interesting cases. With a little encouragement from me, Dr. Li keeps working with Mr. Illich during her last year, July 1991 through June 1992, when she is working mostly on one of the inpatient units. She also asks for my help on a research project, looking at substance abuse in our clinic population. She comes to the clinic a few afternoons a week to see patients and work on her project, and we meet perhaps once a month. We mostly focus on her research, her data analyses, her presentation at a conference.

I see Mr. Illich in the waiting room every so often these days, in his wrinkled suit, poring over dense instruction manuals.

"So what's up with him?" I ask.

Dr. Li tells me that he has finished one set of computer courses and is enrolled in a more advanced course, trying to get certified as a Microsoft technician. His wife, having dealt with her own depression after her mother's death, has gone back to work as a pharmacist's assistant. His kids are doing well too, one finishing at Stuyvesant High School, another at City College, studying mathematics.

Then I lose track of him. Too much going on: new groups starting up in the clinic, a grant deadline, dozens of new residency applicants to interview. Stuff at home too, life as part of a two-career couple with two little kids.

But, finally, her time at MMC approaches an end, and Dr. Li is busy applying for fellowships (I write her letters of recommendation). I keep reminding her that she has to prepare her patients to be passed on. In June of her fourth year of residency, a few weeks before she is to start her research fellowship at Yale, we meet for a last time to review her caseload. They'll be picked up by an incoming third-year resident, from Albania, who starts working in the clinic in July.

Today Dr. Li wears a clean white coat and a navy suit and white blouse with a large bow—not exactly fashionable, almost an inadvertent 1950s look—and earrings and lipstick that is enthusiastically applied. She looks doctorly, she inspires confidence. She tells me of her plan to move to Connecticut in a few weeks. She is excited about her research fellowship but apprehensive about leaving New York, since she has become comfortable here. Her parents have come to the United States and will be here to see her graduate next week. She would like me to meet them. I tell her that I would be honored.

Really, the transformation is amazing.

It's kind of a funny thing. You might say I've been struggling to teach her *DSM* psychiatry—and that she finally has learned it. Making a diagnosis, applying evidence-based treatments and all that, medication *and* therapy, and that somehow, after much work, we have gotten the model to work, even across such vast divides. But it clearly goes beyond that, which is good, really good. It's something that we learned at Payne Whitney, the art of building a "therapeutic alliance," connecting empathically with your patient across such divides. For her, stories build that bridge.

Toward the end, I ask her about Mr. Illich.

Dr. Li smiles. "He surprised me this morning. He wore a new suit too! I asked him why, and he said he got a job."

"A job? It's been years since he worked."

"Maybe eight years. He finished computer training course. Out of the forty students in class, three students were offered a job right away, and he is one of the three."

"How's his health?" I ask.

Mr. Illich is hardly cured, Dr. Li informs me: he still suffers from angina and gastritis, but his headaches are no longer incapacitating since she added the new anticonvulsant Depakote to his medication regimen. But he has become active, energetic, even happy.

"Of course, 'One swallow does not a summer make.'"

"What does that mean?" asks Dr. Li.

I explain that his depression could still relapse. He might get fired. Long term, he's still a suicide risk.

"In China, maybe we are optimistic," she says. "There is a poem from Sung Dynasty, and it says if you see one plum blossom on the branch, it means a whole tree will be blossoming, that spring is coming. That is what I told Mr. Illich."

"Interesting," I say. "That's a beautiful thought."

CHAPTER 7

The Red Box

Digging Deep Into the *DSM*, Late 1990s

At Manhattan Medical Center, in the 1980s and '90s, the "red box" is an actual metal box, not red but 1950s-file-cabinet gray. It holds perhaps a hundred four-by-six-inch index cards with names, dates of birth, medical record numbers, and a summary of offenses. Sometimes eight-and-a-half-by-eleven-inch pages are stuffed inside, detailing particularly risky cases.

The first thing to know about the red box is that it does not exist. No hospital surveyor, insurance reviewer, or city official will ever discover it on their rounds of the psychiatric hospital or during their accreditation reviews. It will never be sitting on a counter in the admissions office or atop a desk in the walk-in clinic, because for all intents and purposes there is no such thing. It is the equivalent of a psychiatric asylum's back ward, in miniature—a concentration of hopeless cases, a distillation of failure to cure—and since such wards no longer exist, neither does the red box.

The second thing to know about the red box is that every psychiatric facility has one. (Indeed, no hospital can survive without it.) It may be called the blacklist, the do-not-admit registry, or even euphemistically referred to as the sign-out file. At one hospital it is a multipage handwritten list kept on the chief resident's clipboard, checked before each contemplated admission. At another hospital years later, it takes the form of a shared Word document on a secure network drive. And as mentioned, at MMC it is a gray metal box.

The third thing to know about the red box is that no matter what else it may contain (phone numbers of ambulette companies, lists of hospitals willing to accept late-night transfers), its true purpose is a warning: a list of individuals whose admission to the psychiatric hospital—inpatient or day treatment or clinic—is to be avoided at all costs.

Membership in the "do not admit" list is not easily earned. Yelling, cursing, throwing objects, even assaulting hospital staff or security guards will not entitle one to admission to the red box. Feigning psychosis to get "three hots and a cot" won't do it either.

On the whole, we psychiatrists are immensely tolerant of strange behavior. We want to believe our patients, even when we know we shouldn't. Almost deliberately gullible, we step into the pitch, so to speak, with regard to credulity. So even the patient who repeatedly comes to the ER to extort handfuls of Valiums or who repeatedly checks into the psych ward when his welfare check runs out—or to escape life on the street—only rarely do these folks get added to the red box. To be added to this select roster there must be a history of repeated abuse, exploitation, attacks (especially attempted rapes of other patients), harassment, wanton destruction, smuggling of weapons and abusable drugs, and the like.

But once a name enters the red box, it's almost impossible to remove it. That person has burned his bridges irretrievably. And the system, so much abused, will do whatever it can to keep him there. So, the fourth thing to know about the red box is that its inhabitants are admitted for life.

Such is the case for the patient I will call Viv R. Her story is told in three stapled index cards among the hundred or so in the MMC red box.

> Name: Viv R
>
> Med Record #: XX XXX XXXX XX
>
> DOB: XX/XX/195X
>
> Dx: chronic paranoid schizophrenia
>
> Course: 1st psychotic episode, soph. year CCNY—hosp @ Bellevue t/f S. Beach. Readm X3 1977 @ Bellevue, t/f Manhattan State, shelter. Rikers Island, MMC St. Vincents Bellevue Hospitaliz. 1979X3, Mar, May, July 1980, ref Day Tx, assaulted other pt, discharged from Day Tx, inappropriate referral per Dr. T.L. Adm MMC 1980–1986, signed out AMA X11. Rx. CPZ, Haldol, Prolix PO then IM/Decanoate. Readmitted again @ MMC X19 btw 1988–1993, refused IM, assaulted sleeping pt/delusional. Tx over Obj. X4, assaulted court app'ted atty. Bellevue, Rikers, Bowery Women's Shelter 1993,4,5.
>
> NOTE: DO NOT ADMIT INPT/OUTPT/DAY PROGRAM

Viv R is a sad case, as is abundantly clear from these well-thumbed, coffee-stained cards, to which is also stapled a folded photocopied summary from an inpatient admission. Apparently, Viv became psychotic when she was in college and since then has been in and out of hospital, in one emergency room after another, sometimes homeless, other times crashing in her father's apartment, frequently arrested for disorderly conduct.

I become aware of Viv R when I am appointed director of the outpatient division at MMC in the early nineties, overseeing the clinic, the day program, and the psychiatric emergency room. For years, Viv R haunts our emergency room. For years, dozens of doctors and psychologists and social workers try to help before finally, reluctantly, red-boxing her. Viv has a pale face and crooked teeth, and when you talk to her, she stares toward a corner of the room, never making eye contact. She is always accompanied by her father. She wears a hat, no matter the weather, and he sports a grimy raincoat, no matter the weather.

She is the one with labels, innumerable grim *DSM* diagnoses; he is the chorus, the escort, the relentless and tireless advocate. Remarkably belligerent, always cursing, complaining, raising a ruckus, Viv has never responded to treatment of any kind. She has evolved a routine, however. Once admitted to hospital, she immediately submits a sign-out letter and refuses everything.

> I want to leave this hellhole immediately.
> I refuse to take any medicines!
> You have no right to give me Haldol.
> Let me go home now!

Her sign-out letters trigger a legal dance: either the unit must discharge her within seventy-two hours or go to court for treatment "over objection." Generally, they cave and send her home. Even on those few occasions when Viv is given medication involuntarily—shots of Haldol or Prolixin—there is no indication her frenzy can be contained. She hits and bites and spits her way through any waiting room or ward. Everywhere, Viv creates chaos: she frightens other inpatients until *they* submit sign-out letters.

Whenever Viv is sent to my clinic for follow-up, she shreds magazines in the waiting room, scaring the kids. She'll show up at the walk-in office, and I'll find our psychiatrist arguing with her and her father, two security guards warily behind them. "Viv, you have to leave!" And security escorts her back onto the street. She'll wrangle another intake slot from a temp worker at the front desk and then be a no-show. After a few weeks, her desperate father will begin to call demanding another slot ("Urgent! Urgent!"), which she is guaranteed to skip. Though sane, Viv's father is no better behaved: he berates innocent clerks, demands to speak to administrators all the way up the ladder, then parks himself in the chairman's office and refuses to leave.

Day treatment, self-help meetings, partial hospitalization programs—everything has been tried, at enormous cost, I suppose, and utter futility. Sometimes she accedes to long-acting antipsychotic medicine, which is injected every two weeks and can keep psychosis at bay. Though she responds less to voices and takes better care of her grooming, she remains suspicious, agitated, and spiteful. After two or three shots, invariably she refuses more. She attends the day program for exactly one day, then disappears. Even when escorted by her father, she slips out after a few hours. In our clinic waiting room, she screams at a woman with three children, threatening them with a brick pulled from her purse, before security drags her away. A few times she ends up in long-stay state hospitals, the place of last resort for such cases, but even they find ways to discharge her in a matter of days.

But nothing can keep Viv R and her Dad away from the emergency room. Perhaps three or four times a week they pop in and insist on being seen by the triage nurse. And the whole cycle begins again.

You get the idea. Hence Viv has the dubious distinction of being the exceptional case used in teaching residents. *Alas, our much-vaunted diagnostic system, based on the* DSM, *now in its fourth edition, revised, namely the* DSMIV-TR, *cannot help everyone.*

"You know the woman with a hat who is always in the ER? Whose father keeps paging you?"

The residents sigh: dad is infamous too, often rallying the page operator to torture them numerous times in a single night.

"Virtually everyone else you treat this year *will* respond to treatment. Our diagnoses are decent, our drugs aren't perfect, but they will help most folks. She's one of the few . . ."

We shrug. It is impossible not to pity Viv and her father. But nothing can be done to help them. Hence, she has amply earned placement in the red box.

One day, I ask Rebeca in the file room to pull her entire chart, trying to find some secret, some way to help. There are five thick volumes in all, kept in a milk crate that sits on the floor of my office for weeks before I dare to look into them. Starting from recent notes, I flip backward from box to box, but it's the same story over and over. Finally, frustrated, I skip to the earliest notes from a dozen years ago. There I find an onion-skin carbon copy of an early hospital discharge describing her "onset of illness."

So, there *was* a before, a Viv who lived a normal life. She is sketched briefly: a sophomore at CUNY, an only child, gifted at mathematics and music. Her father works for the NY Transit Authority; her mother is a nurse. They live in a small apartment in an outer borough, and she commutes to school, first to high school at Bronx Science, then to the City College campus located near Times Square.

She breezes through the first few years of undergrad, studying linear algebra, differential equations, and number theory and playing piano until something happens. She begins missing assignments, skipping classes and lessons; she complains another student is harassing her. Then she disappears. Her frantic parents file a missing person report; eventually she turns up at Bronx State, one of the public hospitals, having been admitted as a Jane Doe. She is kept there for several months and given antipsychotic medicines.

I find thick packets of discharge summaries from nearly every public hospital in the New York area. One describes how she reports being raped three times, though these allegations are neither confirmed nor disconfirmed by what seems like a cursory investigation. Later summaries document how, once sent home, Viv quickly becomes psychotic again, then spends two years on a back ward at another state facility. She is discharged with plans to attend a day treatment program, where she no-shows and is quickly rehospitalized. Again, again, again. Easily a dozen other summaries show earnest attempts at providing continuation care. Which brings us up to today: an endless cycle of admissions to various institutions in the city, fruitless referrals to day programs and clinics and residences.

There's an old snapshot of her as a City College freshman: an unsmiling young woman with pale skin, blue eyes, a dash of inexpertly applied lipstick. Wearing a black high-buttoned dress, her tightly combed hair parted in the middle, she resembles the daughter from the Addams Family whose name I can never recall. Even in that early picture, Viv R is not looking at the camera but instead off to the side.

So much for that. I drop the ragged volumes back into the crate and lug them to the record room. Whereupon they disappear: when I try to find them again, the clerk denies having ever checked them back in.

Is Viv R unique? I doubt it. I suppose that if you look deeply into the case of every Viv you pass on a park bench seated beside a possession-stuffed grocery cart, or begging on the subway, or haunting the median strip of Broadway uptown near Columbia University, clutching ragged plastic bags, or standing with hand held out at the door to your bank, you will find a similar story: a life more or less on track, then suddenly and mysteriously derailed. Indeed, the New York subway and the late-night sidewalks are an open-air *DSM* of treatment failures, of people you can all too easily imagine in their earlier days, at high school track meets or hanging with friends at the mall or surrounded by happy family members as they visit Disney World or the Grand Canyon or beginning army service or excitedly starting a job at Citibank.

The medical New Year comes July 1. Then, as all doctors know, medical students turn into interns, interns turn into residents, residents into fellows, and fellows start real life as practicing MDs. According to folklore and some hard data, it is risky to get sick the last weeks of June or the first few weeks of July, and no practicing doctor would voluntarily admit him- or herself to a hospital at that time. During those weeks, as caseloads are handed off from senior to more junior doctors, there is a good chance for things to fall through the cracks, for immutable rules to be broken, for institutional memory to fail. Even with the best efforts of all concerned, it is a witching hour.

And I believe that the early-summer medical New Year is the reason for what next happens. After years of being confined to the red box, and thereby not allowed admission to our hospital, Viv disappears. Residents on call no longer report being harassed by Viv or her dad; she is nowhere to be seen in the emergency room or the walk-in clinic. I have to admit I am not giving her much thought: too much on my hands with managed care, cutbacks, being a dad, and trying to do my share of parenting and housekeeping.

Has Viv been admitted to a long-term care facility? Have she and her father packed up and left town? Is she haunting another hospital's emergency room, perhaps in a city far, far away? Could she have died?

There are different theories, but all are wrong. Months after her disappearance, I am signing a stack of treatment summaries for Dr. Lamkovich, a third-year resident who recently transferred from another program when his wife matched at MMC's surgery residency program. His regular supervisor is out on emergency medical leave. Halfway through an inch-thick stack of papers, Viv R's name pops out. According to Lamkovich's note, one Viv R is a member of one of the twice-a-week evening outpatient therapy groups, which she attends faithfully, and is taking college mathematics courses.

There must be some mistake. Perhaps there is a *second* Viv R, a doppelganger, a namesake, a distant cousin? I summon Dr. Lamkovich to my office. An Eastern European immigrant, a Chicagoan transplanted to New York, this resident is not easily perturbed. "Yes, I noticed from the computer that Miss Viv had many previous visits. Although she was in my Clozapine group for several months, at first her chart was missing. Finally, they found it behind boxes in the copy room. When it came, I saw it was very thick."

"And?"

"By that time, she was better. When I saw that she was red-boxed, I did not know what to do. So—I have kept her in my group."

"Is this all true? She comes regularly to the group meetings? She participates? She is thinking of getting her own apartment? She's taking *classes*?"

"Yes, she comes to the Clozapine group." Dr. Lamkovich explains patiently, as though to a small child: clearly, he doesn't see what all the fuss was about. "One meeting each week is for the blood drawing and to talk about medicine effects and side effects. The second meeting is to get the prescription and to talk about their lives. Viv is one of the best members of the group. She is also a very good student in her classes. She is thinking of going back for her bachelor degree."

To put it mildly, I'm astonished. Of course, on one level there is a simple explanation: Viv has been prescribed Clozapine, a recently introduced antipsychotic medicine, which works differently from the first generation of antipsychotics, the phenothiazines and butyrophenones like Thorazine and Haldol. Clozapine was first synthesized in the 1950s and was studied in Europe in the mid-1960s; after its release in Switzerland in 1972 and Germany in 1975, it was linked with patient deaths, caused by effects on white blood cell production, so-called agranulocytosis. It was then withdrawn from the marketplace before being gradually reintroduced in the 1980s. In the United States, it was approved for treatment-resistant schizophrenia in 1989, requiring regular blood monitoring. Like all antipsychotics, Clozapine has effects on the brain's dopamine system. It also binds to the brain's $5HT_{2A}$ receptors, working as an "antagonist" to those receptors and thereby helping not only psychotic symptoms but also depression, anxiety, and what are called "negative symptoms" of apathy and withdrawal.

Of all the drugs available to treat schizophrenia, Clozapine is still the most powerful, even decades later, though also potentially the most dangerous. Because Clozapine can cause life-threatening agranulocytosis, or loss of the body's ability to produce white blood cells, which can be fatal, it is only doled out in weekly prescriptions. Patients need to have a regular monitoring of blood counts, starting every week, eventually perhaps once a month. Because of its risks, most doctors will only give Clozapine to patients who are likely to keep weekly clinic visits—leading to a paradox: the exact people who might benefit most from it, whose lives are ruled by chaos, by strange voices, by bizarre thoughts, are generally not thought to be reliable enough to qualify for Clozapine.

The fact that Viv has even been given Clozapine is an artifact of the witching hour of the medical New Year. Face it, Viv is probably the worst Clozapine candidate any of us at MMC has ever encountered. No one—no one who had ever met Viv or her father—could have imagined her to be likely to attend twice-a-week groups or to comply with weekly blood sticks. Only because of an odd convergence of stars—Dr. Lamkovich's transfer to our program in his third year, which means he never took call in the ER and thus never crossed paths with Viv and her father, *and* the missing chart, *and* that she was registered by a temp worker at our front desk, not to mention the utter failure of our red box—has Viv ever been given a chance to get Clozapine. And she has had a remarkable response.

"And her father?" I ask.

"Once Viv quieted down, so did he."

Now he is just another elderly dad who peacefully sits in the waiting room, his overcoat folded on the chair beside him, waiting until the group ends, and afterward he accompanies her through the park. Though actually, most times Viv comes to the clinic by herself.

Once I become aware of Viv's reemergence, I keep seeing her in the waiting room, late in the afternoons, sitting quietly or chatting with other Clozapine group patients. I see her off and on over the coming months as Dr. Lamkovich runs his Clozapine group, which he also elects to continue as a senior resident. I get to know him as a very good, compassionate MD, utterly unflappable. During this time, Viv is accepted back at CUNY to finish getting her BS degree, and once she graduates, she plans to apply for a master's degree program. Eventually she gets her own studio apartment. Oh, and she adopts a three-legged cat.

And so Viv's story ends. Years later, long after I leave MMC and long after her father has died, she is still doing well.

I am sure there is a moral here. Perhaps there are several. One image in particular keeps coming back to me. In one of our conversations, Dr. Lamkovich tells me of his own family background and how his immigrant father worked in the steel business in the industrial Midwest. He got into buying shut-down steel mills in Rust Belt states, Michigan, Ohio, Pennsylvania, for their scrap steel, eventually specializing in buying old steel mill cauldrons. Such cauldrons, so-called ladles, are made of metals with high melting points, aluminum oxide, magnesium oxide, and the like, and lined with ceramic bricks so they won't corrode or melt themselves. It seems that these old cauldrons, in which tons of iron ore are melted for many decades, invariably accumulate a layer of sludge at the bottom, which contains heavy metals; a layer that the elder Lamkovich discovered often contained gold, silver, platinum—all the heavy rare metals. The cauldron sludge is worth a fortune.

Maybe it's a poor metaphor, but it sticks with me: the cauldron and the red box. The cases stuck in the red box, in a way, are the distillate of failure, those who don't spontaneously improve, whose symptoms are untouched by conventional *DSM*-based treatments, whether compassion or Haldol. In a sense, the *DSM* diagnoses are too general—these people fit into broadly defined standard diagnoses, and we have no way of separating them from more easily treated people based on their symptoms or life histories. There's something unique about them, it increasingly seems to me: they too are precious metals. For diagnosticians looking to refine our *DSM* categories, for clinicians who want to try pioneering treatment approaches, for researchers looking for genes that cause mental illnesses, what could be better than hopeless cases?

And many years later, decades in fact, genetics researchers *do* begin looking at the back wards of state hospitals, seeking among such hopeless cases—and finding people with treatable autoimmune diseases and with rare genetic mutations, so-called copy number variations, who may respond to state-of-the-art individualized treatments including gene surgery. I propose they should start searching the hospital and clinic red boxes as well.

In all, Viv's story is one of miraculous happenstance. Clozapine seemed to have a remarkable, life-altering benefit in her case. But why? Why does Clozapine help some people with schizophrenia for whom the traditional antipsychotics don't work? Possibly those folks don't have the abnormal dopamine function seen in most chronic psychoses, or maybe their symptoms result from an abnormality in their brains' glutamate systems, an entirely different neurotransmitter. Also, it is worth noting that schizophrenia is a strange illness, at times waxing and waning on its own: maybe Viv was already going into a phase of relative calm and cooperativeness before Lamkovich welcomed her to his Clozapine group. It's also even possible that we had her diagnosis wrong all along: sometimes people with bipolar disorder are wrongly diagnosed as schizophrenic, and Clozapine can have transformative effects in bipolar as well.

But there is one thing for sure: Viv's story makes me think about who else may be hidden in the red box, at my own hospital and elsewhere. After all, the red box is potentially a source of miracles rather than misery. Who else might escape its terrible oblivion?

CHAPTER 8

Call

Testing the *DSM* Off Hours, 1998

A dreamscape, entering the liquid city before dawn on an August weekend. No traffic on the Bruckner Expressway, the FDR Drive eerily empty except for the odd security guard coming off a shift and a befuddled Nebraska caravan waggling across middle lanes. I even get a spot right in front of the hospital, where ambulettes usually double-park, idling at midday. I am so early. I drop into my office to check emails, then swing up to the inpatient service, where three of the four doctors on call, the DOCs, crowd the residents' office.

Number Four is nowhere to be found.

The incoming senior DOC, a sardonic Romanian named Florin, says he'll page him. Florin's breakfast, a soggy tuna fish sandwich, rests on the grimy desk. It is a typical resident office, where they see patients during their inpatient rotation: coffee-stained carpeting, painted-shut windows, unruly piles of lab printouts and out-of-place drug company Frisbees and squeezable brains from some long-ago conference.

We wait while Florin pages DOC Number Four, the incoming junior DOC. I am the attending on call this weekend, or the AOC, along with two other DOCs. Sana, the outgoing senior DOC, is an adenoidal Indian woman I've worked with since her first year, and the outgoing junior DOC, Jazz, is a lanky young Californian with tattoos and four port-side earrings.

A few minutes later, Number Four blows in, streaming aftershave. Barker is his name. He is lean, pink-cheeked, and bright-eyed as a tympani player in a Midwestern marching band. How did he ever end up at MMC, I wonder, among our hipsters and expatriates?

"Stopped by the ER on my way here, what a mistake!" He scans the room, his white coat crisp and fresh. "So, how does it look?"

"Beds are everywhere!" Florian says.

Sana interrupts. "I don't usually tell the consults, but this may be problem."

Mrs. Drummond, as I will call her, a fifty-two-year-old widow, was admitted to the medicine service by her internist. Five weeks ago, for reasons known only to herself, she plain stopped eating. Her weight is down to 80 percent of normal.

"She needed intravenous hydration. And once her pressure comes back up, her son wants to sign her out. The floor calls me last night 10 p.m.—they plan to discharge today."

"And?"

"She will die. If she keeps on starving herself, she will last only a few weeks. Problem is, her internist agrees to send home. *He's* in denial about the seriousness."

"What about transferring her to Psychiatry?"

"She won't sign!"

"And she's not 2PC-able?" That is, admissible against her will.

Sana shakes her head. "Only if is below 75 percent normal body weight."

I look at DOC Number Four. "Well, she's your problem now, you can figure this out."

He gives us all a bright Midwestern smile.

Which is why I'm here, I guess. I am this weekend's on call attending, thereby completing a two-week cycle. Twice a year, each of us attending psychiatrists carries the on-call beeper. Actually this time I'm taking a third cycle; covering for Thom Singletary, who has been out sick for weeks and whom I'm worried about. The last time we had lunch in the sun-speckled back garden of the KK Restaurant, me and him and Madden, Thom hadn't eaten much, just pushed his pierogis around the plate, smearing apple sauce and pickled onions. What's wrong, we asked him. Nothing, nothing, he said, I'm fine. He almost fell on our way out of the restaurant. Then he went on sick leave. I was happy to cover for him, the least I could do.

Call didn't used to mean much. The residents never, ever called—respecting your sleep time, perhaps, or, more likely, from their own macho pride. The thing being, when Monday came around, especially after long weekends, disasters had often festered for days. So eventually the director clamped down. The proximate reason was the Bell Commission Report, resulting from a tragedy, the 1984 death of Libby Zion.

Libby Zion was admitted late at night to New York Hospital and treated by inexperienced and overworked first- and second-year residents. This poor girl—a Bennington College freshman—had been taking the monoamine oxidase inhibitor (MAOI) phenelzine for depression and was given Demerol by a young resident on call, a classic mistake. Demerol is well known to interact in a potentially lethal way with MAOIs, but her young doctors were apparently unaware of this. She got agitated, was put in restraints, was then given an antipsychotic drug, compounding the problem, her temperature shooting up to 107 degrees. Which led to cardiac arrest and death.

Libby Zion's father, the eminent *New York Times* journalist Sidney Zion, filed a series of lawsuits against her doctors and the hospital. He also started a campaign to limit the number of hours residents could work—at that time, they often worked well over one hundred per week. The New York State health commissioner, David Axelrod, set up an Ad Hoc Advisory Committee on Emergency Services, commonly known as the Bell Commission after its chairman, Dr. Bertrand Bell, to investigate this. Starting in 1989, resident work hours were limited to eighty hours per week, with no more than twenty-four hours in a row. The Bell Commission also set new requirements for coverage of hospitals on weekends and holidays by licensed physicians, that is, attendings (rules that applied to New York but were soon adopted by other states).

Now carrying the call beeper isn't enough; we attendings must actually go in on weekends and round with the residents. We see each new admission, examine them, and write a note. Generally, this means we give our blessing to what the residents have already decided and do our best to ensure that nothing goes wrong on their shift. It's a good thing too, no complaints here. But from the residents' point of view, it cramps their style.

Never mind the lecture hall and the anatomy dissections and morning rounds—every young medical student and resident knows how important call is. Nights and weekends are when the real action happens in hospitals, when decisions are made and lives saved or lost. Call is where students become doctors.

In my own life, call began at an early age, long before I entered medical school. When I was a kid, my Dad, in his months "on-service," would often get called down to the hospital to help the cardiology fellows. Not infrequently, he would bring me or one of my five brothers and sisters with him. It was a central part of our childhood, driving down Cedar Hill to University Hospital late at night, parking in the doctors' lot, and going up to the medical floors to see a patient in trouble, whom the residents and fellows on call couldn't stabilize. Wearing an oversized borrowed white coat, which dragged on the floor behind us, we'd follow Dad along gleaming corridors to the Coronary Care Unit.

There, white-coated doctors would crowd intently around Dad, reporting problems and dilemmas. We would watch them all, lit by the green oscilloscopes, as Dad flipped through the chart, unrolling long waxy EKG tapes, and began decoding the mysteries that were baffling them. "What's going on here," Dad would ask. "What do you see on the rhythm strip? What's the source of this arrhythmia?"

After what seemed forever, saying, "Let's go see the patient," Dad would lead the troops away. Generally, we were left at the nurses' station, but starting around junior high, we would be asked along too, and—feeling like impostors—joined the white-coated brigade at the bedside, where some ashen elderly man was propped up, gasping for breath.

What happened next, sometimes over a few minutes, other times after interminable waiting, felt like being at the heart of medicine, the ineluctable struggle for healing. Dad would talk with everyone, pushing them to come up with a life-saving plan. They would push IV medicines, change settings on the monitors, and sometimes we would wait there, for hours it seemed, to see what would happen; other times we would keep rounding on other patients. Either way, we would eventually come back up the hill and pull into the driveway and come back into the house where everyone else was sleeping, at times victorious, at other times, in crushing defeat, at least knowing that we had done the best we could.

Call was a matter of dread and yearning, one of the things that drew me to medicine—wanting to be part of those life-saving moments—even when having my gravest doubts. And my siblings too: probably no coincidence that four of the six of us became MDs. Life didn't seem complete without it.

Such was the case for me in becoming a doctor. Being on the front lines, finally, after all those years of college and preclinical med school classes, one had the feeling of trench warfare. There were always new admissions, bleeding or arresting or seizing or groaning in pain. Something was always needed *stat*. Example: a young mechanic who had taken a long slide off his motorcycle on a California highway. In the darkness of the Santa Clara Valley, we spent the better part of the night stitching him back together on his hospital bed since no OR was available, suturing huge flaps of skin and fat that had been ripped away, showing live red muscle—until his body looked like a hastily repaired tent.

Even in lulls there was the hospital as a set, a twenty-four-hour theater throbbing with possibilities. The interminable corridors, the stale sandwiches grabbed from vending machines and an unending supply of two a.m. pizza; the squawking beepers and monitors, the scrawled scut lists that necessitated going on aching feet from one bed to another, the central line trays and the chest tube setups wrapped as carefully in sterile blue cloth as holy scrolls . . . Such is the stuff of every doctor's origin story.

But by the mid-to-late 1990s my life is far less glamorous, both during workdays and when I take call. Days, I spend much of my time—like thousands of hospital doctors—training young residents, reviewing cases, and making sure documentation is up to par. I am part of "administration" now. Sure, our work is crucial for saving lives—after all, making sure every doctor and nurse's hands are clean prevents many more deaths than late-night emergency angioplasties—but it's not exactly dramatic. Sometimes it is like being at Mission Control. Like a grounded space traveler, I watch my students, and I fill with vicarious and wistful pride as they rocket above, leaving vapor trails behind them, disappearing into blackness.

Now, being assigned to take call again, it brings the feeling back. Not quite the same as when I was wearing the crumpled scrubs; it's more like being the line coach than a halfback, close by the action but apart from it.

Despite the loss of glamour, I reassure myself: It's okay. It's necessary and good. After all, one cannot be a student forever. And this is where so much healing occurs—in hushed conversations between doctors in hospital wards, on rounds, in on-call offices.

Call also brings up the *DSM*, in a funny way, its glories and its limits. There is no shortage of critics of psychiatry these days—sociologists, philosophers, medical historians, and the like, to mention only the mainstream ones. They decry the "medicalization" of postindustrial suffering; they claim psychiatric authority has expanded into domains of normal existence. It's because of the oppressiveness of capitalism, they say; madness is a result of social inequality and poverty. We overdiagnose schizophrenia in people who are actually depressed, especially among people of color. Okay, definitely, I agree. Or maybe our patients are a type of political prisoner, and we are abusing social power by labeling them as "mentally ill" and forcing treatments on them. Our diagnoses are arbitrary, value laden, full of moral judgment. We overprescribe poisonous drugs to enrich drug companies.

But I've never heard anyone say, "My daughter's suicidal, call a sociologist!"

Call makes you acutely aware of the relatively primitive state of the art, how there is ample truth to many of these critiques of the *DSM* and its related treatments—but also how it's basically the best we've got.

Call makes you choose from the possible. And none of these critiques expand possibilities.

Standing outside the quiet room where a man is pounding his head against the wall, you wish you had more to offer than a shot of IM Haldol for his acute psychotic reaction, but you're damn glad you have it. Let's ask Thomas Szasz to help out, or Michel Foucault. Hand them the order book; let's see if they can do better!

Plus, it's a relief to step away from the daily grind of hospital deficits and cutbacks. Call is simple, call is elegant: just keep them alive until morning, until the next shift comes on.

"The problem, it is legal more than psychiatric. Or medical," says Sana, the senior DOC, referring to Mrs. Drummond. "So frustrating. She is probably highly treatable, if only we could get her agreement."

I nod. "What's your thinking about her diagnosis?"

"Most likely she has severe major depression. Probably depression with psychotic features."

"Psychotic why?" asks Barker.

"People do not starve themselves for no reason," says Sana. I ask her to expound.

Psychotic major depression is a biological illness, she says, that, perhaps surprisingly, responds very well to treatment. Because it is so severe, it is counterintuitive that any treatment can make the symptoms nearly vanish. A combination of medicines (antidepressant and antipsychotic medicine) can be effective or, even better, electroconvulsive treatment. ECT is quicker: sometimes two or three treatments will bring the person back to him- or herself. Generally, given the options available in modern psychiatry, we can successfully treat almost everyone with psychotic depression.

"But you need to have a willing patient," Florin adds. Or a patient who can be legally treated over objection. Neither of which applies to Mrs. Drummond.

"Excellent," I say. "Anyone have thoughts about what we can do?"

Silence.

I point to Number Four, in his well-pressed white coat. "Give it some thought, okay? Seems likely to play out today. Call me if you need help."

"Sure," he says unconvincingly.

Next, we head out on walk rounds on the locked unit. Sana shows us around.

There's a seventeen-year-old girl who overdosed on ibuprofen after her boyfriend cheated on her. A Central Park West widow who convinced the ER staff that she needed a break from her desperate life that only the inpatient psychiatric ward could provide. And a Latino man with gang tattoos who can't hide a smirk when he says he is *desdichado*. Why try to con us? Florin thinks he must need to get off the street for a few days.

And there is a follow-up from last weekend's call, Fazli, a thirtyish man. Fazli's uncle brought him to the United States two years ago. In Pakistan he had been in one mental hospital after another, abused and ignored. Once Fazli got to the United States, he had stayed in a single-room-occupancy hotel, getting worse and worse until the old man dragged him to the hospital.

Last Saturday when we rounded he was locked behind a door with a small reinforced-glass window. He was crouching on the floor, naked, using his fingers to eat out of a plastic cereal bowl. He'd been drinking his own urine, the nurse told us. We maxed his antipsychotic dose right away.

Today, only a week later, Fazli is amazingly better, no doubt the result of the full-court press we started last weekend. Routine admission lab tests showed that he had severe thyrotoxicosis, which had apparently caused a chronic "organic" psychotic episode with agitation, confusion, paranoia, disorientation. It had remained undiagnosed for all those years back home. In *DSM* terms, he has a "Psychotic Disorder due to a General Medical Condition."

With help from the endocrinology consult, facilitated by an Urdu-speaking translator, Fazli is being treated with thyroid-blocking drugs and is being evaluated for thyroid radiation or possible thyroid gland removal. His psychosis is being treated with a new atypical antipsychotic, plus the mood stabilizer Depakote. A parasitology workup is underway.

Now, dressed in crisply ironed hospital pajamas, Fazli strolls around the unit arm in arm with Buyani, his attendant. He smiles in wordless gratitude.

What about Mrs. Drummond, the starving woman? By convention, we attendings don't see consultation-liaison psychiatry cases on the medical floors—the patients who have started pulling their tubes out, or are threatening suicide, or are hearing voices. They are managed by the junior psychiatry residents on their own. I know DOC Number Four will see it as a rebuke if we are to break protocol and do so, so I will never see the patient I am most worried about. Instead, the senior DOC and I strategize with Barker about how to handle things: what to do if she insists on signing out today?

And that is it. After I interview each patient briefly, we stand in the nurses' station, conversing. Then the residents rush off to order labs, write their notes, and see what's happening in the ER. I write my notes and head home.

Barker seems less concerned about Mrs. Drummond than his colleagues. Perhaps it is Midwestern naïveté, I muse, or maybe he just isn't as sharp of a resident as Sana. Whatever the cause, as we discuss Mrs. Drummond's case, he seems almost blasé, as if to say, "Let that lady do what she wants. If she wants to kill herself, who are we to stop her?" Even though, as we all agree, she is an atypical case for self-starvation.

Yet scarcely an hour after I leave the hospital DOC Number Four is ringing me up on my cell phone, sounding scared. "Mrs. Drummond's son is here. He wants to take his mother home now. Now what do I do?"

Not only am I back in suburbia, I'm in the midst of a vast Price Club warehouse, pushing a shopping cart piled with oversized cartons of pasta sauce, soup, and canned fruit. Cell reception is spotty; I roll down the long aisles looking for more transmission bars.

"Technically she has the right to leave," I say. "But I think you should try to play for time. Say she can't leave until she starts eating."

There is a pause that can't be attributed to a weak signal. Eventually Barker mumbles, "And what will that accomplish?"

I ignore his question. "Look, just tell her, unless she starts eating, she can't go home." It is an out-and-out lie but worth trying. My guess is that even if we did transfer her to Psychiatry over objection, it wouldn't help much. If she goes to court to fight involuntary commitment, which she has the right to do, a judge will spring her immediately.

DOC Number Four isn't finished: "And if she really insists on leaving?"

I think for a moment. "I would refuse to let her go unless her internist comes in to see her. Let *him* sign her out. Let him take responsibility for her life." The odds of him coming in today are nil.

Number Four says reluctantly, "Okay, okay, I'll try that." He calls again after dinner. He sounds exhausted. He tried my line, he says, that she had to eat in order to leave. They haven't been able to reach Mrs. Drummond's doctor, despite calling his service, paging him, and so on. "The son, he got tired of waiting and went home. He says, okay the mother will stay tonight, but he is insisting to take her home tomorrow. He'll call a lawyer, he says he'll sue us if we don't let her go. Can I ask you, sir," he adds plaintively, "What's the point of all this? We're bluffing."

"Sometimes you bend the rules."

"For what end?"

I don't know how to answer that question. I am hardly optimistic. "We have to do what we can," I say finally.

So maybe the critics have it partly right. Psychiatry as social control. Manipulating, bluffing, delaying, even outright lying. I'm sure the sociologists and philosophers, the Szaszes and Foucaults and the like, would be gloating. But they're not the ones taking call, with all that it entails.

The next morning, Sunday, is a variation on Saturday. Again, there is no traffic on the drive, and again I slide easily into doctors' parking, to meet today's three DOCs, who are packed into the grimy inpatient office. Today, Florin, the outgoing senior DOC, has an egg salad sandwich before him, plus a huge overflowing Dunkin' Donuts iced coffee. The two incoming residents, Zenn and Kiplander, are here. Again, there is no sign of DOC Number Four.

"Sorry, sorry, sorry, late again!" Barker rushes into the room, his white coat now crumpled and stained, his cheeks blue with stubble. He too carries an enormous Dunkin' Donuts iced coffee, which he immediately proceeds to spill. He mops it up with blue Chux wipes from the nurses' station.

"So how does it look?" asks Zenn, the new junior resident, who is coming on to relieve Barker, a fresh-faced blond woman with a shaved left temple, running shoes, and a sixties peasant dress. She dodges the streams of coffee that Number Four halfheartedly dabs.

"Beds everywhere! We only had two admissions."

Zenn groans. She is soft-faced and looks to be maybe eighteen and a half, but I know she has a PhD in biochemistry.

Barker details the new admissions. "An Irish construction worker—an extremely severe alcoholic, name of Shea, and Mrs. something or other, a paranoid pharmacist from Greece." He rushes through the case presentations, and then everyone gets up, ready to go over to the floor.

"What about our lady?" I ask.

"Nothing new." He last saw Mrs. Drummond at eight p.m.; nothing had changed. He describes her to Zenn, who listens quietly, with a puppy-like air. Will this young kid be able to stand up to Mrs. Drummond's son any better than Barker? I have my doubts.

The new patients are in the other building. We walk outside. Three fire trucks have pulled up in front of the hospital entrance, and half a dozen firefighters holding hoses and axes stand in the lobby, having commandeered the elevators. So, we go around to the avenue entrance and up to the sixth floor. The only way to get to the psych ward from here is to go through Pediatrics, one of the hospital units that is being closed.

"They'll never let us through," whispers the new junior as we approach.

The Peds nurses look up, surprised.

"There's a fire," I say. "They've cut off the elevators, we need to walk through."

They wave us through the nearly empty ward.

Over on Psychiatry, we find a ruddy-faced man with two shiners. "Both of my parents died of drink," he tells us with merciless clarity, describing his own certain demise. "I know how it's going for me. Already, my liver is shot. I know I canna anymore get plastered, only I can't stop!"

The pharmacist is an ancient crone who, it turns out, is only thirty-eight years old. In broken English she explains how five of her neighbors have keys to her apartment and enter it whenever she leaves, stealing her valuables.

After rounds, I write my notes. I review Mrs. Drummond one last time with DOC Number Four and the incoming junior resident. "Look," I say, "I know you're off duty now, but do me a favor, go see her, would you?"

Bell Commission rules: he has done his twenty-four and signed out to the incoming crew; he is legally mandated to stop work.

"Okay, Sir." He manages a big Midwestern smile, and then it is time for me to leave.

Outside, it is still quiet. In front of the hospital, where ambulettes are usually pulled up, idling, and where Hasids from Williamsburg are always sitting in battered Pontiacs, waiting for their bewigged wives, the street is practically empty. A Latino family walks slowly toward me, kids and grandma and parents, dressed for church.

I am just pulling out of my parking space when I get paged. It is Barker, sounding excited. "Sir!" he cries. "I need to tell you! Mrs. Drummond's son is here! He wants to take the mother back home!" His voice crackles through my cell phone. I pull back to the curb.

"You know what we talked about," I say.

"No, no, I think she can leave. Her internist has already come by. She has an appointment to see a psychiatrist tomorrow. And guess what?" She started eating, he tells me. Last night, after he left the floor, she had a full dinner. Chicken, peas, mashed potatoes, ice cream! I hear the incoming junior DOC in the background: something about an egg white omelet, home fries, fruit salad, coffee. "And this morning, breakfast as well."

I listen to him, trying to dissect the truth from his exhaustion.

"She swears she'll keep it up!" he cries.

People will say anything, do anything, to get out of the hospital. Or to sign out at the end of call, I guess. But I wonder: is there perhaps some hope for Mrs. Drummond? No magic here, no curative therapy or drugs, nothing high tech yet in our advancing specialty. Only the possibility of driving a wedge, even with empty threats, hoping to change her trajectory. Modest goals, at best. But at the end of the day, essential.

There is a feeling at the end of call that is difficult to describe. Relief, of course—and fatigue, even if you have gotten a good night of sleep, since you've been carrying a heavy load of possible disaster. But satisfaction as well, at least when you have reason to believe that you have taken care of today's problems and perhaps prevented a few catastrophes. I can remember, after Dad had stabilized the patients in the Coronary Care Unit, the feeling of driving back up Cedar Hill, and when we came home into the dark, still house, where Mom and my five brothers and sisters were sleeping, and when we sat in the breakfast room with its map of the world on one wall, snacking on fried bologna with ketchup on rye toast, or boiling some hot dogs sliced end to end, and the feeling that we would have, or the feeling of Dad's that I would share. You've tested the limits of your approach in the real world; you've done what you can with the tools you have. *Something good had been done. Someone would live.*

And so, perhaps, with Mrs. Drummond, at least for today.

CHAPTER 9

Less with Less

Stripping the *DSM* to the Essentials or Beyond, 1998–2000

MEMORIAL DAY, 1998

"Doc, doc, I just wanted to mention something!"

Dr. Moss catches me at my car as I'm about to head off for the long weekend.

"Nothing to worry about," he adds, approaching warily. "It's about Mrs. Greer, she wasn't in treatment yet."

I set my backpack on the hood and listen. Now I *am* worried. Behind Moss, the main building of Manhattan Medical Center has a strangely derelict look, and not just because of the upcoming holiday weekend. In recent months, it has consistently looked this way, as if hastily evacuated in anticipation of some flood or terrorist threat. Piles of refuse bags accumulate in hallways, rain beats against wheelchairs abandoned in the back alley, dust swirls in stairwells, security posts go vacant. And locked doors have signs with arrows: "Use Other Entrance,"—but the other entrance is locked.

Pale and frazzled, his tie askew, Moss too seems to be part of the general dereliction. Despite his dedication, always notable and at times almost noble, something is often missing in Jack Moss's work, a line forward, an ability to get to the point. Moss tends toward the roundabout most when stressed.

Moss repeats that Mrs. Greer wasn't in treatment yet. She was awaiting her first appointment and had been discharged from the hospital last month. "We don't have to submit the Report of Death."

I gradually apprehend what he is getting to: "Another patient died?"

"The outreach worker," he continues, "went to visit her, and she was making bat noises. High-pitched squealing noises. Very strange noises, and . . ."

Focus, Moss. "So—what did he do?"

"Called Mobile Crisis, who went to her apartment, and *they* called the police, who broke the door down. Unfortunately, she had already hanged herself."

Just two weeks ago, Thom died. We went to his funeral, Madden and I. It was uptown, on the East Side, at the Frank E. Campbell Funeral Chapel. His family was there, sitting in front, tall, Black, silent, somehow unapproachable. Those of us from MMC sat behind them for an interminable time; no one said anything, no prayers or speeches, just an hour of silence. Finally, a thin man with dreadlocks came into the room and began playing the piano. When he was done, we came up to the open casket to say goodbye to Thom, seeing his face, the wasting nearly hidden by the mortician's art, until finally, still silent, we left.

Tonight, away for the holiday weekend at the beach, surrounded by family, I go to sleep thinking about Thom and about how when we asked him at the noodle place if he had AIDS he said no. Madden and I talked afterward, still not understanding why he kept us at a distance after all these years.

And I fall asleep thinking of Moss's patient, Mrs. Greer, and I waken from a dream of falling bodies—bodies dropping through the sky. Recently this has become a familiar dream and one that will haunt me for years to come, long after thousands of other bodies fall to earth just a few miles away.

I lie there, restless in the damp heat, grieving in my heart for my friend Thom Singletary and for Mrs. Greer. I can hear the pounding of surf, the faint rise of drunk lovers going at it on the boardwalk to the beach.

> I hate you! I haaaaaate you!
> Fuck you fuck you!
> Fuck you!

Mrs. Greer should never have been let out of the hospital. Shouldn't have had to wait three weeks for an appointment. Shouldn't have had some undersupervised outreach worker, no doubt some twenty-two-year-old college kid, standing outside her door, confused by the squealing and thrashing inside. There was no need to summon Mobile Crisis—anyone experienced would have called the cops immediately.

And Thom, Thom, we wish we could have helped you or that you even could have let us know when you needed help.

Mrs. Greer could have been helped. It wasn't so difficult. A forty-seven-year-old woman who lived alone, Ms. Greer was diagnosed with what the *DSM* would define as undifferentiated schizophrenia. Having responded well to a course of the new antipsychotic risperidone, following two weeks of hospitalization, she needed to be treated in an outpatient setting, preferably a "partial hospital program" where she could be seen four or five times a week for six to eight hours a day. But all the partial programs have been closed, so she was referred to a local clinic in Greenwich Village. They were having staffing cuts, so couldn't see her right away. The next appointment was in a month.

We failed her, just like we have failed half a dozen others who have died since winter. We know what tools would help her; they're just not available. These days, there's no slack for the Mrs. Greers of our catchment area. It's not cost effective to keep them alive.

"A forty, maybe fifty-million-dollar deficit at Manhattan Medical Center this year," our director announced early this past spring in the executive committee meeting. "Maybe one hundred million next year. Things aren't likely to improve in the foreseeable future." We—the senior attending docs—sat in stunned silence. A hospital-wide deficit of ten or twelve million would not have surprised us. But fifty million was immense.

Within a day or two, rumors flew. It didn't help, as *Crain's Business Weekly* reported, that Moody's had downgraded MMC's bonds to junk. Things were so bad, we heard, that the hospital president's limo driver had been laid off. And the Pediatrics Department—*Pediatrics!*—might be shut down. At our uptown sister hospital, part of the same health care system, Dentistry was abruptly closed. "And it wasn't even losing money!" they say.

In Psychiatry, I knew immediately that my outpatient division was vulnerable. We would bear the brunt of the cuts. Our programs might be excellent, our staff amazing. We are doing fantastic *DSM*-inspired psychiatric work. Makes no difference. It is not a matter of trimming and scrimping—I've done that innumerable times. I would have to close something, most likely our day treatment program. And let go most of its staff. And soon enough, we closed day treatment. We were able to squeeze one MD into my clinic and a few social workers into vacancies on the inpatient ward, but its nurses and other staff were soon told they would be fired. One worker, Elaine, came by my office in tears. Just married, and pregnant, her husband unemployed, Elaine had no idea what she would do. Disconsolate, she just sat there, as if waiting long enough would bring back her job.

Even on the inpatient service, morale was tested. A few days into the layoffs, a psychotic girl was assaulted on the ward by a streetwise addict, also a patient, in the middle of the day. No staff members were around in time to intervene.

And so, over the wet months of spring 1998, as we approach the end of a decade, a century, a millennium, our normally bustling hospital has taken on a sense of siege. In the hallways, we doctors who always stand before elevator banks quietly consulting about patients, or rushing through the park on our way to the floors, now become silent with worry, pale with rage. We are tired of the endless excruciating meetings where we are berated for our refusal to cut muscle, since there is no longer any fat left to cut. We are exhausted from trying to do more with less, from trying to provide care for the masses of sick and injured and dying with fewer doctors and nurses and techs. We ache, in continual irresolvable mourning. What is happening to our beloved hospital, which has been serving the Lower East Side for over one hundred years? How can we be abandoning our patients?

It is not just New York, we soon realize. It is a national thing. A shakeup—or shakedown—of the health care system. *Disruption*, like what banks and investment firms have been going through since the early 1980s. Some people say that word with glee, as if it is a good thing to shake things up, make everything more efficient, leaner, with less redundancy, to strip everything to the basics and then strip some more. Except with us, it is not only vice presidents and analysts being laid off: patients' lives are at stake.

When I return to the clinic on Tuesday, after the long weekend, there are already about a dozen patients lined up at the registration desk. A lone security guard barricades the walk-in clinic, attempting to keep several "holds"—patients being held for observation—from escaping. Outside my office door stands a resident with prescription pad in hand, desperate for my cosignature. "All prescriptions must be stamped or imprinted with the name and medical license number of a licensed physician," says the recently updated New York State Prescribing Law. The residents generally don't have their New York State medical licenses yet. So, every five minutes all day, five days a week, comes a knock at the door: *Sign, stamp, sign, stamp.* A constant syncopation.

Soon, our ER doctor calls from his cell phone, letting me know he's stuck in traffic on the FDR Drive, so the walk-ins will have to wait. One of my docs pokes his head in: Transportation won't bring in his suicidal patient without "administrative approval." That's Franklin's job. Franklin, my administrator, has been out with colitis (no doubt stress related) for weeks. So, I call Transportation and argue with them to send a goddamn ambulette to Avenue D.

It is a fair amount crazier than a few years ago, with a lot less teaching since two of our senior doctors were "bought out" (that is, involuntarily retired) in January, each with over thirty years' experience in supervising residents. On top of that, I've had to fire two MDs, one social worker, and two part-time psychologists. Because of the austerity budget, several other lines are being held vacant.

Today, like the beginning of every week, I'm irrationally hopeful. I always arrive at work convinced that there's got to be a way to make it work. It can't be as impossible as it seemed last week. Somehow the weekend has assured me yet again that the state of the hospital in which I left work last Friday was accidental, a meaningless blip.

Most urgent are the overdue reports to New York State on two patient deaths. One who jumped off the roof of his building a week after discharge from the hospital. Another who turned a pistol on herself rather than on her abusive husband. The reports are weeks overdue. It used to be that we filed one or two Patient Deaths and Serious Incidents per year. In the first five months of this year, we've added six.

HISTORY

When I think back on those days, Manhattan Medical Center, amazingly enough, appeared to be one of the healthiest hospitals in town. We were among the likeliest to survive.

For decades, MMC was the prototypical sleepy local community hospital, albeit in the midst of a huge city. It grew gradually in the first years after I arrived, adding new departments, programs, studies, but it remained local, patched together, a friendly and intimate little place. Then, perhaps a decade after my arrival, things began to change. Initially, the pace was gradual: a nursing home acquired, a few group practices absorbed, a failing community hospital handing over its buildings and dissolving its board of trustees. While grander and more prestigious hospitals embarked upon building campaigns, erecting huge inpatient pavilions with glassy atria and grand granite lobbies filled with postmodern sculptures and gleaming fountains, when it came to the megalevel of development, "the vision thing," MMC's trustees seemed unable to get it together. One master plan after another was presented at medical staff meetings, each vast new imaginary edifice touted as the answer to our needs. Yet nothing was built.

Docs who worked at the big teaching hospitals in town would mock those of us who were stuck at MMC, with our tangled corridors, our low ceilings, our warrens of ancient brick dispensaries and limestone-clad nurses' residences, our creaky elevators. We were doing great work, sure, but in a crazily outmoded facility.

The only thing that our board seemed able to get off the ground was a big new ambulatory care center, half a mile from the hospital, in space recently abandoned by some financial firm gone bust after the latest Wall Street crash. And even that was just a field of dreams—echoing halls, a gloomy atrium, floor after floor of abandoned trading facilities filled with huge computer monitors, dusty high-end ergonomic furniture, and exposed pipes and ductwork, like a set from Terry Gilliam's film *Brazil.*

Would MMC's initiatives be enough to cope with the wildly changing world of medicine? If you looked beyond the local scene, it was obvious that the health care industry was exploding at the seams. Medical costs had ballooned from 5 percent of the U.S. budget in 1960 to 8.9 percent by 1980 and to over 12 percent by 1990. American health care costs were easily double those in Canada, Europe, and Great Britain.

Overwhelmed, near bankruptcy, the U.S. Medicare system began a system of "prospective payments" based on patients' diagnoses, so rather than charging for every service that was rendered, hospitals were reimbursed a fixed amount for each "diagnostic related group," or DRG. These fixed payments were intended to motivate hospitals to make care more efficient. Then came a system of "relative value units," or RVUs, which were intended to find a common currency to value the benefit of all medical treatments. Comparing the value of treating, say, a case of acne versus a case of depression, or metastatic lung cancer, or a heart attack. Acne cryotherapy (freezing zits), for instance, went at 0.77 RVUs a pop, so to speak; intubating a patient to put her on a respirator was worth 2.33 RVUs; and a psychopharmacology visit yielded 1.29 RVUs. RVUs had some effect, but not enough by far. Soon, employers began to try to control costs by setting up different types of management structures, each with its own acronym: the HMO, the PPO, the physician hospital organization. These also had little effect on ballooning health care costs.

Then, in the mid-nineties, managed care appears. Perhaps the most innovative and perverse new organizations are the managed care organizations, or MCOs. The geniuses behind MCOs figure that the best way to cap medical costs is to "capitate": to pay doctors and hospitals a fixed amount of money per member of an organization each month to provide care. In some ways this makes sense: if you have $1,000 per member to pay for medical care, then the doctors and hospitals need to find a way to provide it *within* that $1,000. Capitation incentivizes doctors and hospitals and clinics to reorganize, to do their best with the money available.

It becomes quickly clear that there is more money to be made—*real* money—by *not* providing care. Why spend the entire $1,000 on treating the patient? Why not find a provider or clinic or hospital able to do it for $800, or $500, or $200, and pocket the rest?

We see it coming in the mid-nineties. We hear from our friends in California and Minnesota that something big is on its way, an avalanche, a tidal wave, something that will sweep away everything in its path. Our friends at the big academic centers in town scoff at managed care. They have big new hospital buildings; they provide state-of-the-art tertiary care. They always have more patients than they know what to do with, always have, always will. As the top brands, they can charge more for the services they provide. With such leverage, what do they have to fear?

Then, hospital bed occupancy rates begin to fall. I start getting calls from my friends at the fancier medical centers, newly laid off, asking if I know of any jobs. The university hospitals with their costly new buildings have suddenly realized that they are stuck with white elephants, with enormous fixed costs that they might never cover. This is when everyone begins to say how lucky we are over at MMC.

So, strange as it may seem, despite our more-than-occasional disillusionment, the morale at Manhattan Med is probably higher than at most places. After all, we still have our jobs. The new party line is this: Manhattan Medical Center could be a winner in the new age of medicine. Merely by being so far behind, we might be ahead of the curve.

The chaos we live amid has its very obvious destructive side but also spurs innovation. We spend our creative energies—when not staving off disaster—by continuing to investigate innumerable topics raised by the *DSM-III* and by developing new evidence-based mental health treatments. Our work in psychiatry at MMC is a huge hit. Except we can't possibly treat all the people who need care, and we are being overwhelmed by the emerging demands for treatment of these disorders—and by increased costs and falling reimbursement. What we face in psychiatry is echoed by similar problems in internal medicine, in pediatrics, in geriatric medicine, based on a host of new, life-prolonging, and often expensive treatments. Better treatments, leading to yet more demand for care. Which butts up against the enormous inefficiencies and waste of the American health care "system."

But there must be some way to make it work. We keep innovating: more group therapy, hiring per-diem therapists for our evening and weekend clinic, increasing productivity. Maybe group therapy is the way to go: treat more people per hour of therapist time. The therapy groups we set up to deal with panic disorder, women's issues, trauma, AIDS psychiatry, and dual diagnosis (for people with psychiatric disorders and addiction)—all have grown into separate programs and draw patients from all over the tristate area. We keep getting more research grants; we cut deals with managed care organizations to treat employees of local companies. Here at MMC, we are all used to doing the impossible. And so, for a number of years we have convinced ourselves that the pain we all feel is that of rebirth and that our anguish is that of unwanted but inevitable metamorphosis.

REENGINEERING

"Hey, how's it going?" says Eva when I run across her one Monday, a few weeks after Labor Day, 1998. We've met on the street, rushing to our departmental conference.

I shrug, pleased to see a friendly face. Eva is one of the inpatient docs, and we've worked together for a decade. "Surviving," I say as we wait for the elevators. "What about you?"

Eva scowls. "They've drafted me for the next phase of reengineering. We've got to cut two million from the inpatient budget."

Of the three elevators, two are out of service. The sole working one seems to have gotten lost in its shaft.

"Sorry to hear that."

"Well, they can't hurt me anymore."

"Two million sounds like a lot of pain."

"No," explains Eva. "I've accepted that there's absolutely nothing else that they can take away."

We wait interminably, and finally go for the stairs. Breathless from the long climb, we arrive on eight, the administrative floor, with the first fall executive meeting well underway. We take seats at the back of the room. Our director is speaking. Once kind and shyly genial, now he is inward, brooding, unpredictably veering between enthusiasm and dismay.

Weirdly, he is saying: "Great news!" He lays it out, surprising us all. In keeping with the trend across America, our hospital is going to merge with another hospital in the city. "This is going to be a big step forward. I see great things coming from this!"

Merger. I look around the room at the other docs—my colleagues and rivals and friends—to gauge their responses. Once we're merged, explains Dr. Cadmus, there'll be more leverage with managed care companies, bigger market share. More patients (as if my clinic needed more patients!).

"We'll be more competitive than ever."

Next, Madden, who is now medical director of the Inpatient Division, begins to explain reengineering, a key part of the merger. Inpatient units are being reengineered first, he says.

The stresses of the past few years have wrought great changes in Madden. A fitness nut, a weightlifter and open-water swimmer, in recent months Madden has acquired a permanent pallor on his once-tanned face and a scowl that suggests an ongoing dyspepsia.

These days everyone asks, "What's going on with Madden?" He's a good doctor, always has been, but really he excels at teaching young medical students and residents; over the decades he has been an inspiration for hundreds of doctors who practice throughout the country. He has an edge, he loves to dish, he always has something snide on his tongue. These days, though, Madden doesn't gossip, he doesn't complain. He doesn't even call to yell at me anymore about "inappropriate admissions." He just hides in his office.

In a monotone, Madden fills us in. Until now his main goal on Inpatient has been to get unskilled—that is, cheaper—workers to replace more expensive professionals. "What a nurse did a few years ago, now we assign to a nursing aide, what a psychologist did, now we give to a 'psych tech,'" he explains. "As much as possible, to use students and interns and externs—who don't cost *anything*—to take the place of paid staff."

I watch Eva's face: as far as I can tell, she's ascended to a Zen peak of invulnerability. You can sense serenity in her gawky frame and dispassion in her broad forehead.

"But now we've got a problem," Madden continues. "Basically, there's nowhere else to make cuts, without seriously compromising patient . . ."

Cadmus stops him mid-sentence. "Of course, we won't compromise patient care!"

Lunch arrives, a tray of deli sandwiches and a bin of soda cans, one of our few remaining perks. We dig in. Before break, though, Cadmus has one more announcement. "Because of our uncertain financial picture," he says, "the three-per-cent raise you were promised for January is being postponed indefinitely."

Eva is hunched over pastrami and rye. I see it on her face: Cadmus has hurt her after all.

LOST IN THE ARCTIC

Our strategy in dealing with the onslaught of managed care—the only possible one, the only one leading to survival—soon becomes clear. We need to cut *and* to grow. To make everyone work harder, do more with less. That is the only hope of erasing our deficit.

It is strange, what is happening: in so many ways, it resembles disease. A system that is ailing, with something awry, is pushed beyond its limits. Staggering forward, continuing as usual at the risk of catastrophic failure—until a hip gives way, angina turns into myocardial infarction, psychic stress results in a breakthrough of suicidality. Or is it more like a hurricane, a natural disaster, far beyond the individual efforts of any of my doctors or social workers or psychologists, beyond the heroic efforts of our director or the hospital administrators? Perhaps even beyond the scope of the greedy bosses of the managed care companies. Economic winds—like the drought that created the Dust Bowl during the thirties. Or if not entirely natural, still inevitable, like the struggles of Western farmers in the late 1800s to keep the family farm afloat, fighting with banks to keep their lands, local catastrophes resulting from global financial forces. An impossible task, which didn't prevent them—or us—from trying.

Soon after our reengineering meeting, Annette, probably our most dedicated doctor, comes into my office. We've always kept her incredibly busy: teaching residents, running groups, supervising evaluations. Now she has two little boys, her husband has been downsized, and we're saying she needs to "increase her productivity" by 40 percent. She starts calmly, but soon tears begin sliding down her cheeks. "I'm sorry, I just can't do it!" she says. "I've met my match."

Annette gives us a month's or so notice, but already she is gone.

Soon it becomes clear that the situation is dire: even hospital vice presidents are getting axed. To cope, they send us to management seminars to learn buzzwords—things never taught in medical school.

Continuous quality improvement. Downsizing. Rightsizing. Vertical reintegration. Throat-sticking jargon. It's like we manufacture shoes or coats in some outmoded factory and are being threatened, thanks to NAFTA, by the remorseless forces of market capitalism, by cheaper offshore workers.

One day, we docs have a mandatory "management retreat" in a Midtown office building, where we are each given a vinyl binder with lots of tabs. Internists, pediatricians, gynecologists, psychiatrists, pathologists are gathered in one room—the whole spectrum of MDs from MMC. Our instructions: "You'll learn to cooperate by pretending you are lost in the Arctic. Teams need to collaborate in order to survive."

We do the exercise faithfully, but one team wins by a landslide. A surgeon on that team, it turns out, already did this exercise at some *other* management retreat. A ringer. He tipped his team members to hang on to their maple syrup container (metal, can be used for cooking) and warned that they'd die if they started trekking across the tundra.

We are hardly alone. All over the city, all over the country, hospitals are struggling for survival, trying to do more with less, many of them going under. Locally, Long Island College Hospital, Interfaith, Cabrini, North General Hospital, St. Clare's—all are sinking. Over in the West Village, St. Vincent's Hospital is practically bankrupt, enraging local residents by its plans to sell off its valuable real estate.

Everywhere, large consulting firms are having a field day, advising hospitals how to close beds and eliminate services. And corporate raiders are capitalizing on the perverse incentives of capitation, making tens, even hundreds of millions of dollars by denying care. Insurance company CEOs enjoy seven- and eight-figure bonuses, inflated by such shenanigans.

At MMC, the Inpatient Department might be shrinking, decreasing the number of beds, losing its clientele, but Outpatient continues to explode. We have hundreds more patients, thousands more visits, but less staff and fewer resources than ever. Desperate for additional income, we accelerate our deal making. With Family Medicine. Primary Care. Local clinics. Social agencies. Everyone expects that we will treat *their* patients as number one, with minimal wait times, fast turnaround, whatever services they request. *Sure, your patients come first, no problem.*

Rather than two or three thousand visits a month, suddenly we are overwhelmed by four, even five thousand per month. Our excellence, our amazing work, does not go unnoticed. Our successfully treated patients refer their husbands and aunts; they bring their kids; their GPs refer their depressed diabetics and cancer patients; local colleges send their students in crisis for help. And there is no shortage of crises. AIDS, HIV, homelessness, opiate abuse, suicidal adolescents . . . we are there for everyone with our *DSM*-inspired treatments. Our patients are sicker and needier than ever, medically and psychiatrically ill, addicted, on the brink of homelessness. It is like a fierce nighttime battle with no historian to record its drama and greatness. I beg our director to let us hire another full-time doc to supervise and see patients. But he refuses: no money in the budget. "Do your best with what you have."

Why are we suddenly losing so much money? Despite our massively increased numbers, our grants, our money-making evening clinic? For one thing, we are charged the same overhead per square foot as Inpatient, as Ambulatory Surgery, as Radiology, even though our costs are dramatically lower. No one trusts Finance; no one knows where their numbers come from. The numbers that make me lose sleep, that make us push staff past their limit, that you could say are forcing us to abandon our patients—are they real?

We have ample reason to doubt the hospital's numbers, which have been tweaked to make us share the overhead of more profitable divisions like Surgery and Radiology, but we have no reason to doubt the realness of the *DSM*-diagnosed disorders we treat at MMC. They are undoubtedly real; they profoundly affect the quality of life of our patients, their work functioning, their family relationships; and undoubtedly they increase mortality rates, as I am seeing to my distress at MMC. I see their effects from the perspective of a doctor with fifteen years of experience working in the clinic and of a researcher trying to study the effects of these treatments.

From my perspective, these disorders are certainly as real as the "medical" disorders treated by my fellow doctors in endocrinology and GI clinics. Just like diabetes and cardiac disease, they can often be crippling, even lethal. Depression is real, bipolar disorder is real, schizophrenia is all too real. Borderline personality disorder is complicated but real. Attention deficit disorder is real even though it has fuzzy boundaries, as does depression. Suicide is a devastating consequence of untreated and undertreated disorders. Though our treatments aren't as effective as we want them to be, they can decrease risk of suicide and improve life satisfaction and functioning. We have major problems making treatments available in our society, but that doesn't mean we should give up or throw out our diagnostic systems or treatments. The modern world is plagued by high rates of disorders, which is no surprise given the numerous catastrophic conditions that are present in daily life.

There is a strange disconnect between *DSM* disorders and the brain, of which we psychiatrists of the late 1990s are all too aware. The studies we have as of 1998 show that, somewhat paradoxically, each *DSM* disorder may not be a single "thing," with a single cause or a specific set of brain-circuit abnormalities. A single diagnosis of major depressive disorder or obsessive compulsive disorder may reflect many possible brain changes and may result from numerous causes, but such conditions are still "real" and long lasting, often persisting for years, if not decades. They generally can be effectively treated, where treatment rarely means cure but can mean significant improvement.

Here's another part of the paradox: Even though the people with any single diagnosis may have a wide range of different brain problems, they can often benefit from a small number of treatment approaches. There may be a hundred different types of depression from a thousand different causes, but most people will respond to SSRI medications and cognitive behavioral therapy. Why these treatments are broadly effective in such diverse underlying conditions—none of us really knows.

So if I look at the *DSM* approach and try to evaluate its strengths and weaknesses, I see that it is immensely powerful in the real world. It can help define problems quickly, and it leads to the development of customized treatments that can be tested and shown to be helpful or not. It is scalable to the real world, though subject to competitive marketplace pressures, especially when you have to compete with other medical specialties for scarce resources. And, perhaps not initially intended as a goal of *DSM* psychiatry, I've seen personally how it can have profound antistigma effects, because its approach is so quantitative and so "reliable"—different clinicians can agree that particular diagnoses are present or not. Once they are defined, these disorders can be measured and tracked, and their impact can be quantified, for instance determining the cost of untreated depression to society and its effects on health care utilization. By doing this, it is clear how common and disabling psychiatric disorders are across society and how much they need addressing (that doesn't mean hospital administrators or insurance companies will give us the money, just that we need it!).

The *DSM* approach has a number of weaknesses, as well, I can see. Not only the disconnect between diagnosis and brain dysfunction but other issues. Looking exclusively through the *DSM* lens can lead to a bean-counting approach to psychiatric disorders and thus to the human condition. The *DSM* approach is a "mindless" model *and* a "brainless" model *and* an "atheoretical" approach. Its simplicity, its greatest strength, is also its greatest limitation. It can lead to an extreme of rigidity and narrow-mindedness. While with the *DSM* we can agree (more or less) on what we see, it entirely fails to account for things we cannot easily quantify, which are among the most interesting aspects of the human condition.

It also leads to many complications: with so many different diagnoses, it is impossible to determine by using its methods whether there are any underlying commonalities. Many people "meet criteria" for numerous diagnoses. Depression, panic disorder, social anxiety disorder, avoidant personality disorder—all in one person—can't be four different problems. But what underlying issue or problem or abnormality connects them? On a broader level, are there really five hundred or more different diagnoses, or perhaps only a dozen or so types of common brain problems that can have innumerable different presentations and symptoms?

Also, it is clear that the *DSM*'s proliferation of diagnoses with each new edition has led to a narrowing of the definition of "normality" and to a pathologization of the variability of normal behavior.

Finally, in real-world use, *DSM* psychiatric *practice* has often been reduced in a dangerous way, as is painfully obvious to me at MMC. Psychotherapy has almost been abandoned by many psychiatrists across the United States, who increasingly provide brief sessions, sometimes ten minutes or less, to renew medications. Drive-through psychopharmacology, pill pushing, whatever you might call it, my psychiatrists can be reduced to mere prescription writers. Pushed by the capitalization of psychiatric suffering, *DSM*-based clinical practice can easily be degraded, with an overreliance on cookbook treatments and one-size-fits-all approaches.

So in all the *DSM*, in the Age of the Clinic, has introduced a limited and necessarily flawed model, but it has been a massive step forward. Though it's hard to keep this in mind in the frenzy of work in the MMC clinic.

I am back in my office now. There is a knock on my door: the first of the afternoon's patients whom I must see to fulfill my "practice quota," to help the department offset my salary. I close the door, call-forward my phone, and we begin.

Then comes another knock. "Do you have a minute?"

Sue Vago, MD, one of the third-year psychiatry residents, sticks her head in. Sue is one of the best, selected as chief for next year. I follow her into the hallway, shutting the door so my patient won't hear.

"I just need to know," she whispers, "what forms to fill out when there's a patient death."

"Oh," I say.

She is a thoughtful person, usually unflappable and confident, which is why she will make a good chief resident. "I'm really, really sorry to bother you," she says, following me into the deserted record room.

"No problem," I say. "Who was it?" Sue has the same expression that Annette had that day when she said she couldn't take it anymore. And Elaine, from the day program. Not to mention Jack Moss, when he came to tell me he was leaving to work for a drug company. Which makes my heart ache with pity and helplessness, my face burn with rage.

"I don't know if you'd remember him, he never made any trouble. And that's why I cut him back to every eight weeks. I mean, he was coming once a month for years. Came in, picked up his prescriptions, barely talked to me, so I didn't think he'd mind. And I had so many patients, I just had to . . . had to cut some of them back . . ." Tears flood from her eyes, her face convulses, she forces herself not to wail.

"Look, here's the forms," I say, putting a hand on her shoulder for a moment, feeling like a total shit. It's frustration, anger, helplessness because I'm responsible but without resources. I pity my doctors, our patients, I pity myself for wanting to change things . . . though it beats me how that would be possible. "I'm so sorry, Sue. Just fill them out and leave the chart in my box, I'll look at everything later." And I run back to my patient.

I'm late for the big hospital-wide update about the HSC merger. It is in the grand old marble-clad auditorium in the main hospital building. A standing-room-only whirl of white coats, green scrubs, head caps, jauntily slung stethoscopes, of docs and administrators, rallied from all across MMC, every department and division and center and ward, from Medicine, Surgery, Obstetrics, Pathology, Pediatrics. Only a few years ago I would have recognized everyone here, but today the room is swamped with unfamiliar faces, no doubt from the large cadre of new Family Practitioners and Primary Care docs, who swagger around with the implication that they, as "gatekeepers" of the new managed care system, are already our bosses.

Lights dim, PowerPoints flash above the CEO, who struts with a wireless mic. Everything's going to be great, he tells us. The CEO summons the heads of Primary Care and Family Medicine, our new bosses, to stand up in the audience, and they do, turning around to wave, then he unleashes a blizzard of org charts and facility maps and the like and introduces the CFO. In her sleek big-shouldered power suit and gold jewelry accents, the CFO is ebullient. The merger will be a goldmine, as she tells it. Contracts with huge employers, *hundreds of thousands of covered lives!* A $2.1 billion operating budget, over three thousand hospital beds spread across seven facilities, MMC and HSC will no longer compete. Instead, we will complement one another. It's not so much a marriage of convenience as a union of neighboring empires.

I squirm, watching the audience, which is subdued, its applause polite at best. No one dares to ask anything difficult. Luckily, my beeper sounds. I wriggle through the crowd, out to the grand marble lobby, and hurry back to my clinic.

Dr. Vago has left the neatly completed "Report of Death" and the green patient chart in my box. It belongs to a Mr. Ki. I remember interviewing Mr. Ki at a teaching conference a few years ago. An immigrant from Taiwan, Mr. Ki was too disabled even to wash dishes in his uncle's Chinese restaurant; he was suspicious, secretive, and convinced that there was nothing wrong with him. It was a major struggle just to get him to take his meds. A slim man who regularly heard devils arguing inside the light fixtures. After I interviewed him, I remember, he had shyly asked if he could show us something, and his doctor passed around some drawings. Chinese calligraphy, ink brushed on paper, broad masterful strokes showing striking rock formations, pouncing tigers, wild waterfalls—and interspersed with these traditional images, ballpointed in, hordes of strange robots, tiny Godzillas, rockets, mushroom clouds, insect men with boombox heads.

As per usual with Dr. Vago's work, the chart is impeccable. One note after another describes how Mr. Ki came faithfully every four weeks, picking up scrips for Prolixin and Cogentin. And how, this past January, pushed by *my* push for bigger caseloads, she decided to cut his visits from monthly to once per eight weeks. Ki seemed agreeable; after all, he was merely coming to fetch pills—scrips that could as easily be written out for sixty days as thirty.

Ki kept his first appointment, in March, but missed the one in May. He had no phone, so Dr. Vago mailed a form letter to Ki's hotel (nicely photocopied in the chart), and, busy with her daily activities, did nothing more until Ki's landlord called this morning to let her know that he had died.

> Discussed case w Mr. Wang, landlord; also w Detective R., from local precinct. Am attempting to contact pt's sister in NY. Landlord normally saw Mr. Ki reading newspaper on the street corner. Became concerned after not seeing pt. for two days; came into apt. Found pt. lying naked in full bathtub, plastic bag tied around neck. No note found. Mr. Wang notified police. Det. R. says no evidence foul play. Medical Examiner's review of case pending.

There is a yellow Post-It on the top of the Report of Death, in Dr. Vago's neat handwriting: "IS THIS OK?"

I scrawl: "Paperwork is fine. Please fax to Dept. of Health."

ORIENTATION

After the merger, a frenzy of construction begins. Chalky footprints everywhere, grit on the floors, plastic sheets draped over desks and blocking corridors, and soon walls start coming down. It is quite a project, remaking our "clinic" into a "behavioral health center," retooling us for the age of managed care. Along with an infusion of construction money come new logos, a new letterhead, ad campaigns. Hospital administrators get huge raises and new titles and strut through the hospital sporting their new Ermenegildo Zegna suits and $25,000 gold Rolexes. "They shop," we joke, "and everyone else gets more work."

I gather today's charts together and gingerly make my way through gritty corridors and go to meet the new residents. In the conference room, the new residents and interns calmly wait for their orientation to begin. I introduce myself and welcome them to the Outpatient Service ("which should be one of the most exciting and challenging experiences of your medical training") and remind them to contact their supervisors.

As I speak, I can see the entrancement begin. Young doctors desperately need to learn; they are hungry for initiation into healing. So, temporarily, I exorcise today's reality and try to bring a disappearing world back to life. It is easy to imagine things as they were only three or four years ago, to imagine that these residents will have time to listen to their patients and to talk with their supervisors about the relation between mind and brain; it is as easy as dreaming. And I can do it—the orientation, that is—without much effort, having been through this process so many times before. Then it is time for the assistant director, Dr. Carillo, to do his part. He begins distributing caseload lists.

A few weeks ago, I ran into a former college classmate, a businessman. I was explaining how the venture capitalists had descended on Medicaid. They hit upon giving out stuffed animals and gym bags as incentives for patients to sign up for "managed Medicaid" programs. Trading in a "gold card" level of insurance, that covered everything, unlimited prescriptions, doctor visits, hospitalizations, for new managed Medicaid plans that would ration and limit care. And that would often divert them from their regular doctor to a lower-priced "provider"—a physician's assistant in a faraway clinic with inconvenient hours. Patients were often shocked afterward to realize that because they had unwittingly chosen a new insurance plan, which had its own limited list of doctors and clinics and hospitals, they no longer could continue the treatment they had been getting—whether for their diabetes or their schizophrenia.

My friend shrugged. "Probably the Wall Streeters saw that the system has inefficiencies."

"Many of my favorite patients are inefficiencies," I answered. He shrugged again.

Not a fit subject for today. When it's time for me to speak again, I remind the residents that Dr. Carillo and I will be available to answer any questions, introduce the next speaker, then slip away.

I'm having trouble breathing. Nothing dramatic, just asthma. A bit of wheezing here, shortness of breath there, it's gotten worse over the past year, and the construction dust isn't helping matters. Back in my office, I find that my Ventolin inhaler is empty, so I head over to the hospital pharmacy, itself in the midst of a dusty construction zone.

There I meet my buddy Madden, whom I haven't seen since Thom Singletary's funeral in May. Turns out he's come to pick up medicine for his duodenal ulcer.

"Walking wounded," he says. He starts telling me about the new deal with Medicare and residency training, which will require a 30 percent decrease in the number of specialists. "If this goes through, we'll only have seven instead of ten residents a year."

I try not to wheeze too obviously. Stupid pride, I guess, but I don't want to look weak even to Madden.

"We might have to make one of our units a nonteaching unit," he says. "The words 'nonteaching unit' send chills up my spine."

I know what he means: going from being a teacher and mentor to a plain old service provider, a cog in the machine, is crushing.

The pharmacist slides over brown paper bags containing our meds. I wait until Madden turns the corner before I rip mine open and puff the inhaler.

BOUNTY HUNTERS

Up at the front of the room, next to Dr. Cadmus, sits Dr. Petra, our new managed care expert, who has been brought in to get us contracts from the devil. Sorry, to get us managed care contracts, capitation plans, that kind of thing. Petra is not too popular among those of us who have been here for a while, but we understand why she is here, and we kind of feel sorry for her, given the impossibility of her charge.

Today Petra has a special announcement: "Exciting news!" Our department is going to form its own managed care company, she tells us. We will get paid directly (by someone, I can't keep it straight, either the employer or an insurance company) so many dollars per member per month. A big chunk of money, for which we are "at risk." She's put details on a PowerPoint slide.

Actually, as I squint at it, it's a very small chunk of money. It comes to just over two dollars per member per month, or twenty-four dollars per year per "covered life." I start doing the math in my head: *Even assuming only 10 percent of "covered lives" need psychiatric treatment, the new contract will provide all of $240 for all psychiatric care for that 10 percent!* Fine, that $240 could pay for half a dozen outpatient therapy visits. But what about the members needing hospitalization for depression or psychotic illness, which could cost tens of thousands of dollars per person? Wouldn't that guarantee that you'd lose money? Furthermore, why does the insurance plan only provide two dollars per member per month for mental health? What on earth has happened to the rest of the money? I know from my research that someone, some Wall Street–inspired entrepreneur, no doubt, has skimmed off at least two-thirds of the dollars before they settled on the princely sum of two dollars per member for month for mental health care.

Petra is saying: "Which means that if we spend too much time seeing patients, we will lose money, but if we don't see them much, we can do quite well." As Petra sketches it out, when a patient gets referred to us, we will get one visit to do a full evaluation. We have been used to taking two or three visits, especially with complicated cases involving medical, psychiatric, and social problems.

"The only way to make this work—because we are getting so few dollars per member per month—is to complete the evaluation at the first visit." Then we will request permission for up to six visits. We can get more visits, so long as we send in a request form every month.

"What if we see a patient only once per month? Then we have to send in a request form every single visit?" That was me, stupid enough to ask.

Petra smiles. "Why, I guess that's right."

Dr. Cadmus is staring at me. "I think we'll hold questions for now," he says.

It's strange: I'm not against the *idea* of managed care, and given that I'm in favor of providing health care to all, it seems to me that managed care could be one of the more efficient (and humane) ways to provide it. Our society's health care system is an expensive wreck. Am I focusing too much on possible disasters and too little on the many thousands of people we are helping? Maybe, but that doesn't help me sleep better at night. I need . . . what do I need? I need something different. Maybe some kind of a place where these rules don't apply, where the struggle is to find new treatments for psychiatric illness rather than just to survive—if such a place actually exists.

I just can't accept the moral responsibility anymore. The need for mental health services is too huge. Our treatments help. The *DSM* model—reliable diagnoses, evidence-based treatments—is basically sound, despite its obvious limitations. The problem is societal, a matter of priorities. Not enough money is allocated to do the work. We are being squeezed by implacable market forces. We can't win, at least for now. Maybe these deaths are collateral damage; maybe the system will adjust itself to provide more care for the people who need it, the patients with chronic mental illness; maybe the reimbursement rates will be raised; maybe someone will figure out a way to make it all work.

One more update from Dr. Cadmus: "I want to thank everybody. We have successfully eliminated the two-million-dollar deficit on Inpatient and the one-million-dollar deficit on Outpatient." He also wants us to know that the hospital has just hired a consultant, one of the Big Five accounting firms, to cut costs further. He isn't sure if it will be an additional 10 percent or 15 percent, but the consultant has been offered a bounty based on the amount they are able to cut.

"So, they will find things to cut."

In a way, you could say, that includes me.

EXIT STRATEGY

"You can park on Riverside Drive," he tells me. "Just make sure there's nothing stealable inside your car . . . you might come back to find a broken window. Come in, take the elevator to the sixth floor . . ." I follow his instructions, entering a vast glassy atrium, take an elevator and follow long corridors, and find the glass-enclosed conference room, with an endless view down the Hudson River and a long curved wooden table, where I'm asked to sit at one end, facing the interviewing committee. A dozen Columbia psychiatry professors, asking me about why I want to come to the New York State Psychiatric Institute and what I have accomplished in my years at MMC.

The first job I don't get: that position, for deputy director of the New York State Psychiatric Institute, goes to an insider, a full professor, an internationally known researcher on personality disorders. But then I get a callback, a few months later: do I want to come to interview for the position of clinical director? The same drill, parking on Riverside Drive, entering the airy atrium. It's not entirely new to me here: I did spend a year after residency doing fellowship training at the New York State Psychiatric Institute, an old 1930s yellow brick building with long dim corridors, but since that time they've built this modern blue and green glass edifice that curves along Riverside Drive, overlooking the Hudson River and the George Washington Bridge.

What do I gather from my first visits? Impressions, mostly. They live in a different world here. They aren't being crushed by managed care and impossible work quotas and endless deficits; they are doing cutting-edge research. It seems to be a complicated place, with a huge general hospital across the street, a strong affiliation with Columbia University, and at the core of it a state-funded research facility, the New York State Psychiatric Institute, with dozens upon dozens of famous researchers pursuing their incredible studies. Every type of research, from public health to epidemiology to clinical trials to lab studies of mice and monkeys and Petri dish preparations and EEGs and other biological measures used for psychiatric investigations.

And, regardless of what is happening elsewhere with managed care, somehow, they are expanding rapidly. I come for a second visit and am offered the job. It's intoxicating, even from a momentary view of the place, so much so that I get lost driving home, missing my turn for the Cross-County Parkway, and end up winding along the Hudson River for miles before I figure how to turn east and head toward home. I get the offer letter a week or so later and begin counting the days.

PART III

The Scanner, 2000–2023

CHAPTER 10

Flights Into Health

Learned Safety and the New Neuropsychiatry, 2000–2007

THE TRIP

Hank is a patient from my Manhattan Medical Center days who stays with me when I move uptown to Columbia in the late summer of 2000. When I first began working with him in the late 1990s, Hank was in his twenties, single, harried, underfed, always driven and distracted.

At MMC he would always arrive at sessions sweating and unshaven, his blond hair wild, lugging heavy cases of financial documents, always on deadline, always falling behind. A bookish guy, he was doubly tormented, not only by symptoms but also by the "total mess" he was making of his life. It was hard to argue with that characterization.

A junior associate at an accounting firm, Hank toiled, he slept, he worried, he binged on videos and junk food—that was his life. His existence was regimented and fearful, constrained by invisible walls and barriers, by safe drops and no-go zones. Venturing to my Lower East Side office elicited extreme trepidation, since my neighborhood occupied a boundary between sketchy and outright dangerous—not so much from junkies in the park outside MMC as because of its eight-block distance from home.

Disorder had imposed a painful simplicity on Hank's life. The accounting firm, in a Wall Street canyon, was safe; his apartment, in the Village, also safe; the territories between those places, traversed by cab or on foot (never by subway, which terrified him), were perilous. Nearly everywhere else was intolerably dangerous.

Simply put, severe agoraphobia had decimated Hank's life. Agoraphobia: *an extreme or irrational fear of entering open or crowded places, of leaving one's home, or being in places from which escape is difficult.* In the *DSM*, agoraphobia is one of the anxiety disorders, categorized along with early life separation anxiety, social anxiety disorder, specific phobias, and panic disorder. To be diagnosed with agoraphobia, you need to have marked fear about "two (or more) of the following five situations"—using public transportation, being in open spaces, being in enclosed places, standing in line or being in a crowd, or being outside of the home alone. "The individual fears or avoids these situations because of the thought that escape might be difficult or help might not be available in the event of developing panic-like symptoms." That was Hank to a T.

On his rare days off, Hank stayed home, ordered in, and kept the blinds pulled, since even sunlight and the rush of clouds intolerably assaulted his senses. Should he deviate more than a block from his usual route between home and work, he was crushed by physical anxiety. Ears ringing, heart bursting in his chest—a suffocating terror of death surged until he fled to a safe location, a doorway, a shop, a bank lobby. He knew it was time to get treatment when he began seriously contemplating covering the windows of his apartment with brown butcher paper: having read about people whose condition had deteriorated to this point, he didn't want to end up like the Collier brothers, trapped in piles of debris.

For many months Hank and I worked together with *DSM-IV*-inspired approaches. Having made the diagnosis, I applied evidence-based psychotherapy and medication interventions, using cognitive behavioral therapy techniques and an SSRI antidepressant. First he struggled to stabilize his fears, which were mutating on a weekly basis, and then, when he began to improve, we set the audacious goal of retaking Manhattan. In college at NYU, Hank had loved walking the then-dangerous streets of the Lower East Side; during Columbia graduate school uptown, he dared to venture through Morningside Park, which in those days meant taking one's life in one's hands, and through southern Harlem into the wild woods of northern Central Park. His prior adventurousness only made his current limitations more humiliating.

As treatment proceeded, we measured his progress block by block—a celebration for getting past Twenty-Third Street, the memorable day he dared to take the subway again, and the afternoon he braved weekend crowds in Times Square to buy DVDs at the HMV superstore. Blocking panic attacks with SSRI medicine, combined with disciplined breathing and relaxation exercises, and challenging his catastrophic fears—the campaign to modify Hank's violent internal reactions continued month after month after month.

Over time, Hank reemerged into life as well. He adopted a dog and began taking walks to the local park a few blocks from his apartment, where he forced himself to sit on a bench and do breathing and mindfulness exercises while pretending to read the *Times*. It was at the dog run that he crossed paths with Josie, a lawyer who owned two rambunctious cairn terriers. They dated for several months, then, in an ever-accelerating timeline they quickly married; they moved to a larger apartment and had one child and then another. He found himself able to travel farther afield: after a fair agony, he could visit Josie's parents at their home in western Massachusetts and honeymoon and then take family vacations in the Adirondacks.

Hank was one of those cases that I talk about with colleagues, a success story of the kind you increasingly come across in our field. Thanks to *DSM*-inspired changes in treatment options—making a clear diagnosis and coming up with an evidence-based treatment plan, often combining therapy with behavioral changes and medication treatments—people who have been mired in chronic symptoms can increasingly return to normal life. The key turning point, in my view, is when you get the disorder into a prolonged state of "remission"—so the main symptoms have faded away to the point that the criteria for the *DSM* diagnosis are no longer met. Then life can start to return to normal.

How often does this occur? Is it a rare or a common event? I contend that for the mood and anxiety disorders it is common. In fact, most patients I see with chronic mood and anxiety disorders (my specialty area) can be effectively treated to remission, so their core symptoms are basically much diminished, even gone. In many cases they make gradual progress but struggle to make up what they have lost.

In others, perhaps a third of patients, I see something else, something remarkable: an accelerated trajectory of improvement after decades of suffering from their disorder. They may go through a period of grieving of lost opportunities and wasted years, yes, undoubtedly. But then—as if playing catch-up—they often rapidly move ahead with their education, finishing college or entering graduate school, getting more satisfying work, and building more fulfilling close relationships. Full-time employment with benefits including good insurance coverage is one common outcome for those who had been underemployed because of their disorder. Marriage or committed partnership is another outcome; feeling ready to have kids is yet another. While this kind of outcome is quite common for people with mood and anxiety disorders and for bipolar disorder too, it is far less common for those with schizophrenia, for which we don't have as effective treatments.

Treatment clearly has worked in this case, one muses when facing patients like Hank, but why for him or her when you'll try the same things for other people and not get the same level of results?

Fast forward to 2004. I've been at Columbia for several years, settling into my new job as clinical director of the New York State Psychiatric Institute, and Hank is one of a handful of patients who followed me to my new uptown practice.

It is a beautiful early fall evening when I buzz Hank in. He is tanned and relaxed, clearly in a good mood. It is just after Labor Day, and many of my patients are returning from vacation, ready to face the demands of the fall. Hank seems to be no exception. These days, we meet only occasionally, every three months or so, for med renewals and therapy checkups, during which he fills me in on the latest life developments, ones unimaginable when we first started working together.

"Did I tell you about my trip?" Hank asks today.

"A trip? Where'd you go?" I haven't seen Hank since Memorial Day. At first, I don't recall whether he mentioned that he might be traveling this summer.

"Well, it was my first plane ride, and I went with Josie and the kids to Mexico." It takes me a few moments to put his words together. This is the first plane ride in his life. "It was incredible," he is saying, "To see the turquoise water below us, all the white sand beaches . . ."

Excitedly, he goes on, describing how they rented scooters, went snorkeling, and climbed the steep pyramid at Chichen Itza. Given Hank's lifelong—and all-consuming—fear of traveling (not to mention heights), I had assumed he was going to be telling me about a long weekend in the Berkshires, details of a white-knuckled ride up the Taconic Parkway. But no, he is talking about Mexico. He leans forward in his chair and hands me his mobile phone with tons of pictures—Josie, kids, pyramids, clouds. His suit is wrinkled as ever, his tie askew, his physique a bit more toned, thanks to Josie, and his hair shorter but only a bit less wild than it was when we first met many years ago. But there is an air of confidence about him now, of satisfaction. How did he do it?

Our work over the first few years was plodding, slow, iterative, requiring lots of dialogue in therapy sessions, efforts to push the edges of the envelope that generally resulted in meaningful but incremental progress—a few blocks gained, a driver's license renewed many years after expiration.

But now, apparently, there's been a breakthrough, entirely unanticipated by me and not discussed in therapy. Hank planned this whole trip on his own. He describes how, with Josie's help, he strategized to get through the plane ride: taking an extra half pill of clonazepam that morning, doing diaphragmatic breathing and vigorously "thought-transforming" while the plane sat on the runway.

Once they were aloft, he felt positively giddy with relief and excitement. After all, Hank's entire life was based on the fear of leaving home. In fact, his entire family was marred by this condition. In seventy-seven years, his mother has never once flown: a small-town girl, born on Long Island, she never traveled the fifty miles west to Manhattan until Hank got married. His late father, disabled by a chronic back injury and depression, had rarely ventured out of Suffolk County either. Two of his sisters still lived with their mother. Now, post-Mexico, it is clear to Hank that his all-consuming anxieties had often stopped other forms of exploration as well, blocking many possible improvements in life. Having successfully flown to the Caribbean, Hank can see no limit to what else he might do.

Hank tells me that he is feeling healed, feeling "whole." A feeling that, I wonder that day, as a neuroscience-minded psychiatrist (as I am now beginning to redefine myself), might represent a particular (and for him a new) brain state. As he puts it, "I'm a regular person now." His trip results from—and further strengthens—this wholeness.

THE SAFETY CENTER

Something had changed for me too since I left Manhattan Medical Center in the summer of 2000 and started working uptown. As I settled into my new life at the Psychiatric Institute and Columbia's Department of Psychiatry, I gradually realized that my colleagues and I were embarking on our own giddy adventure—not only the researchers who worked in labs but also the clinical psychiatrists like me, those who worked in offices and hospitals and research clinics.

We were beginning to apply neuroscience advances—knowledge often gained from studies on mice and rats and other laboratory animals—into our office practices, with people suffering from depression and anxiety disorders. After all, it seems clear that the brain systems involved in these human disorders are mostly ancient ones that developed millions of years ago, early in mammalian evolution or even before that. Conversely, as I learn from discussions in conferences and discussion groups at NYSPI, it also seems possible to use observations from patients we see in our offices to come up with new "targets" to study in the research lab or clinic.

As I settle in at NYSPI—pronounced "NISPEA"—I come to realize why it has become one of the best places in the world to investigate these connections. NYSPI is part of a vast health complex in northern Manhattan, an agglomeration including the many buildings of NewYork-Presbyterian Hospital, the Columbia Physicians and Surgeons medical school, and various other research and clinical institutes extending eastward from the edge of the Hudson River to the plazas of Broadway and beyond. The state of New York has provided funds to operate NYSPI as a research institute for over seventy-five years, intertwined with this massive health complex.

Beginning modestly in the 1930s, NYSPI now occupies two large buildings and parts of many others; it is centered around the Kolb and Pardes buildings. Kolb (officially known as the Kolb Annex, located at 168th Street a block west of Ft. Washington Avenue), a tall red-brick building packed with research laboratories, is where the Nobel Prize winner Dr. Eric Kandel and other basic scientists do their research. A lucent elevated bridge above Riverside Drive connects Kolb to a vast modern curved green-glass edifice, the Pardes Building, where I have begun to work—named for the legendary Herbert Pardes MD, who was the chairman of Psychiatry, then the dean of Columbia's Medical School, then president of NewYork-Presbyterian Hospital. Pardes is massive, stretching for several blocks along the Hudson; its southern wing is devoted to work with patients, whether clinical or research, including three hospital units and innumerable clinics and human research laboratories; its northern wing is dedicated to cell and tissue culture labs.

My new life at NYSPI is demanding—and incredibly exciting. My morning starts early with a drive from home in Westchester County to the institute on upper Riverside Drive, just below the George Washington Bridge. I work at NYSPI through the afternoon, doing administrative work and research and teaching; then, on the two or three days I see private patients, I head downtown. From mid-afternoon until well after dark I sit in an elegant high-ceilinged office off Central

Park West, in a suite shared with several other therapists, doing psychotherapy and psychopharmacology, seeing half a dozen or more patients. Nothing special: this is the standard gig for NYSPI faculty—our private practices supplement often-meager hospital salaries.

It is wonderful, at times exhausting, a life of endless too-muchness. Not to mention that my wife, Lisa, has her own more-than-full-time career, and we have three kids to raise, and two dogs, a hundred-year-old house, and long commutes between the city and Westchester County. Once a week I also spend time at a research clinic in midtown Manhattan, seeing patients and supervising clinical trials of antidepressant medications.

I am always coming and going, it seems, but once I arrive at whichever location I might be working, my new life affords me a great sense of calmness. It is such a relief not to feel boxed in anymore, not to be limited by the crush of financial constraints we endured at Manhattan Medical Center. At NYSPI I can focus on the latest scientific progress; when in my office, I focus on the lives of my patients. By this time, I have been seeing psychiatric patients for nearly a quarter of a century; from my MMC years I have accumulated a vast reservoir of experience that resonates with each new issue I encounter as clinical director, and that gives a context to the cutting-edge research conducted at the institute.

And I can begin to make connections, for instance, one day when Dr. Eric Kandel gives grand rounds. A professor in our department, he won the 2000 Nobel Prize in Physiology or Medicine for his work on the neurobiological mechanisms of memory as investigated in the giant neurons of *Aplysia*, a sea slug. Dr. Kandel's presence has permeated the institute for decades. Not only his physical presence—he is a slight, white-haired man who favors red bowties, white shirts, and dark suits and has an elfin, lively manner, and you may encounter him in the building's lobby or walking on the skybridge over Riverside Drive—but his intellectual presence. His passion for connecting the laboratory to the clinic, for translating his studies of *Aplysia* or genetic knockout mice to the workings of the human brain, whether in health or disease, has inspired generations of researchers.

Columbia's Department of Psychiatry's seven hundred faculty members work in all areas of psychiatry, some dedicated to research in a dazzling range of areas, others to patient care, and in daily life we rarely have reason to talk to one another. Grand rounds, though, a weekly conference, is one place where we mix it up: researchers talking about their latest work, in dialogue with clinicians and students from dozens of disciplines. It's a ninety-minute conference, one hour for presentations followed by half an hour of questions and discussion. And it brings top researchers from all over the world to share their ideas with other scientists and with psychiatrists and psychologists who practice in the community. Often the most inspiring speakers are our own faculty members, who have worldwide reputations. None more than Dr. Eric Kandel, one of our two Nobel laureates (the other is Richard Axel, who has done groundbreaking research on the neurology of vision).

So Dr. Kandel's grand rounds is something I don't want to miss. I hurry into the ground-floor auditorium and grab one of the last seats. Our chairman introduces Dr. Kandel. And the lights go down.

The essence of his talk today is this: For a long time, neuroscientists have focused on the brain's "fear center"—the amygdala. The amygdala, named from the Greek word *amygdale*, because it resembles the shape of an almond, is located deep in the brain, in the medial temporal lobe, just in front of the hippocampus. In panic disorder, in PTSD, and most likely in depression as well, the amygdala is hyperactive, hypersensitive—and all too easily sent into "red alert" mode. For over a century, behaviorists like the Russian physiologist Ivan Pavlov have focused on provoking responses from this fear system. In today's version of such experiments, a mouse is put in a cage whose floor is electrically wired to deliver mild shocks. When zapped, a mouse responds by freezing, by hunkering down to the floor and remaining motionless. If a musical tone is played repeatedly just before the shock occurs, after a certain number of repetitions, the mouse learns to associate the tone with the shock.

Dr. Kandel quickly reviews the model of classical fear conditioning: as any psychology 101 student knows, after conditioning, if the tone is played *without* the shock being delivered, the fear-conditioned mouse will freeze, anticipating the worst. Neuroscientists have shown that this punishment model involves activation of the brain's fear center in the amygdala, which increases the production of fear chemicals like adrenaline.

But—Dr. Kandel goes on—there is another side to the fear response system. Recent research in his laboratory conducted by Michael Rogan has shown something novel and surprising: the brain *also* has what might be called a "safety center." This center, located in the dorsal striatum, communicates a sense of safety, of security. If the amygdala sends off air-raid sirens and tells an organism to dive for safety, the dorsal striatum rings out "all clear!" Rogan's ingenious recent mouse experiments demonstrate the powerful impact of this system.

Kandel continues, describing how Rogan brilliantly reenvisions the Pavlovian scenario by asking: What if you condition the mouse by playing a particular tone when there is *not* going to be a shock? Not surprisingly, the mouse soon learns that it is *safe* when this particular tone is sounded—and this tone becomes a clarion call that "everything is OK."

Conditioned safety, not conditioned fear!

Confident that it won't be shocked, the mouse now moves freely around its cage. This "all clear" signal is processed by a different part of the brain—rather than activating the amygdala, it passes through the dorsal striatum's safety center, which connects to parts of the brain that are, as Rogan puts it, "dedicated to the processing of reward, reward contingencies, or positive affective states." Signals of safety lead to the production of brain chemicals like dopamine that enhance the sense of pleasure, even happiness.

Next, Dr. Kandel shows a slide demonstrating how these safety signals affect the mouse's travel patterns. Normally when kept in a cage, a mouse will scurry along its edges. It does not like wide-open spaces, where it will be vulnerable to predators such as owls or foxes. It will hug the wall of the cage, running around the perimeter—only occasionally darting through the middle of the cage. In the wild, such quick forays are probably rewarded with food, increasing the mouse's chance of survival despite the risk of being devoured. With fear conditioning, the mouse won't even do this much traveling—it will hunker down, staying nearly immobile, anticipating the worst.

What happens when the mouse gets the "all clear" signal? Something very different. The mouse now runs into the middle of the cage and stays there, basking in its sense of freedom and safety. It actually *avoids* the "safer" cage edges, preferring to spend most of its time in the usually forbidden open spaces.

As Dr. Kandel speaks, human analogies come into my mind: postgame celebrations at a football game when everyone rushes onto the field, recess period at elementary school, armistice celebrations after a war ends. At these moments, people are just like mice, I realize: when universal signals of safety are broadcast—*often a tone or sound!*—like our furry cousins, we humans flood into usually forbidden zones.

Sitting there in the darkened auditorium, I start thinking about my patients. Over the years, I've seen so many folks with long-standing psychiatric disorders, like Hank, go through a remarkable transformation as they began to recover. During years, even decades of overwhelming symptoms, they hunker down in fear, immobilized like the electrically shocked mice. Even after treatment, as symptoms fade, oftentimes they would remain in one place, still fearing the worst. Just by habit, it seems.

But then, for some patients, as with Hank, something new seems to kick in. At a certain point in treatment, it is not unusual, I realize, for such patients suddenly to go on a trip. Not a harried business trip, some overnighter to Des Moines or Chicago, and not a weekend to fulfill an obligation to Mom back at the homestead. Instead, they seem to book a trip they've always yearned for but that something has prevented them from taking.

Up until now I have thought of such trips as much-needed getaways for people who have been mired in their disorders. In the midst of disorder, when chaos reigns, escape may beckon, but such trips are generally miserable, making a mockery of the concept of vacation. More often, depression or panic disorder make withdrawal essential, make it necessary to hunker down in your apartment or house, sometimes literally to hide under the covers. At a certain phase of recovery, though, such long-yearned-for adventures become possible, enticing, and, as with Hank, irresistible. Now, hearing Dr. Kandel speak, I realize that they might hardly be mere vacations and certainly are not what therapists used to call "flights into health"—desperate attempts to avoid neuroses. Instead, they could be a manifestation of brain *recovery*.

At noon, Dr. Kandel's prepared lecture ends, and the Q&A period begins. I take a deep breath and raise my hand. The chairman nods in my direction, and when it is my turn, one of the chief residents hands me a portable microphone. "Dr. Kandel, that was a really amazing talk," I say, or something like that: I don't recall my exact words. "I'm wondering if we see evidence of the same phenomenon in humans, if we see activation of the safety center for patients with anxiety disorders who have gone into sustained remission. Patients often tell us that they feel whole, they feel like themselves again—they feel *safe!*—and they start exploring the world again."

I have no recollection of Dr. Kandel's answer; I am aware of everyone in the room looking at me, and I flush and my ears ring: probably he says that there isn't yet any human data to bear upon the issue. I do recall that afterward several colleagues from the Depression Evaluation Service congratulate me for my courage in raising my hand and that one of them says, in what I later realize to be high praise at NYSPI: "That's a researchable question."

APPROPRIATELY ADVENTUROUS

For the first time in adult memory, Hank knows what it means to feel safe. But he is hardly the only patient in my practice to describe such experiences. I realize that many others describe something similar once their symptoms fade away.

Take Aline, a young woman who developed severe PTSD after sexual assaults. I see her in my office a few weeks after Dr. Kandel's talk. I can't help but be struck with the parallels between her story and Mark's, not so much the details of suffering as her trajectory of recovery. (An aside: I also can't help but think of Dr. Kandel himself, a childhood refugee from Vienna, a Holocaust survivor, who has essentially metamorphosized into a brilliant scientist and pioneering neuroscientist, training hundreds of students over more than half a century: a remarkable tale of a different sort of recovery. On yet another note, these days I often cross paths with Dr. Kandel at the medical center's pool, located in the medical student dorm next to the Pardes Building, where he swims laps daily; I've wondered whether swimming has contributed to his remarkable resilience.)

The victim of two rapes during her college years, nearly losing her life, Aline struggled with PTSD and major depression for a decade afterward. She became dependent on alcohol and marijuana and lost several jobs. She eventually got abstinent, thanks to regular attendance at Alcoholics Anonymous, and started psychotherapy. While SSRI medications could have helped her depression and PTSD, she refused to take them and instead started an intensive practice of yoga, often two or three or more hours per day. She practices deep breathing, intensive muscle-relaxation exercises, and she has switched from junk food to a Mediterranean diet, including fish and green vegetables and the like. It took over a year, but she was able to go back to work and finish college; she began dating and soon moved in with her partner Jenn, later her wife. She became a yoga teacher, then decided to go back to school. She is now finishing graduate school in social work, with the goal of becoming a counselor, helping other women who have experienced trauma. Over time, Aline too has described feeling "safe" and "whole" for the first time in over a decade. She and Jenn recently traveled nearly halfway around the world.

What I wonder, though: is this just the use of a word—"safe"—or does it reflect something fundamental about how each of these people approaches the world? By this time, I have searched out the safety conditioning papers that Dr. Kandel described in his talk. They are not easy reads: animal studies, dense with technical language, very different from the clinical research papers I usually study. But every so often, a phrase strikes home. In one paper, Michael Rogan describes how, after a period of recovery, of safety conditioning, the mice start a process of "appropriate adventurous exploration."

So too, it seems, have my patients. Hank and Aline: their lives have been dominated by anxiety, by agitation, fear, and melancholy—and now are infused by a sense of calm, relaxation, confidence. Something fundamental has changed for each of them. I can't believe that it is only a matter of their *DSM* disorders being under control or that their psychic conflicts have somehow resolved. It seems like a different state of their brains.

The question being, can our new tools in this age of psychiatry help us understand the essence of such changes? "A researchable question!" The problem being that there is such a gulf between the kind of research I do—with my colleagues in clinical therapeutics—and what goes on in the labs. Our studies in the Depression Evaluation Service—testing medicine versus placebo or one psychotherapy against another—use rating scales of symptoms, whether anxiety or depression or insomnia. Our treatments aim to get *DSM* diagnoses under better control, so if a patient's depression symptoms drop by half or more, they have "responded" to treatment. But we have no idea what is going on in their brains. Some of our studies involve electroencephalograms (EEGs), which measure electrical signals at the surface of the brain. The EEGs might demonstrate an increase in theta waves, for example. How *that* relates to what is going on deep in the brain, however, one can only guess.

"What's changed in their brains?" laughs a colleague in one of our DES discussions. "Nothing!"

"Okay," I retort, "That's the null hypothesis, very useful!"

But even if that is true, what if our current tools aren't powerful enough to prove that such changes *don't* occur? Finding myself in a state of chaotic ferment, I am both entranced by the possibility of studying something meaningful—and overwhelmed by the impossibility of finding any way to get to the brain. There's no point in me going into the lab, by the way, working toward a PhD in neuroscience—not at this phase of life. Anyway, studying animals or cellular cultures has never been my strength; my mind doesn't work that way.

The solution comes one day at work, when I hear an update about NYSPI's new MRI machine, which is now up and running and has additional capacity to add new studies.

When I was lying in the MRI scanner a few months ago, getting a workup for my bum neck, after a tennis-related injury, I first had the thought of adding MRIs to my own depression studies. *What am I waiting for?* I begin reading about MRI imaging and take an introductory MRI course. I can't say that I master the physics involved, but I do understand that by using a huge magnet and an immensely powerful computer you can measure the diffraction of water as affected by the magnet and that you can then use powerful computer programs to construct a three-dimensional image of the brain. Compared to classic X-rays, in which the brain's soft tissue is a hazy smear, the MRI can resolve its structure down to the millimeter. Plus, different kinds of magnetic sequences can be used to study the brain's activity, not just its structure. Different tasks, both cognitive and emotional, can be used to activate different brain centers, since brain activation related to thoughts or perceptions increases the blood flow to specific areas, and the MRI captures this activity.

In an ideal world, I'd want to study safety conditioning, but I will gladly settle for studying other things in the brain—how disorders affect the brain's anatomy and connections and how treatment changes its functioning.

And new neuroscience research shows at least one thing for certain: that the human brain changes throughout our lives, based on our genes and experiences and influenced by the ways we live our lives. Our behavior patterns, our recurring thought patterns, our activities, our diets, our exercise regimens—and especially our lifelong learning of new information—can all change the brain. I keep thinking about the great 1960s movie *The Graduate*, when Mr. McGuire drags young Benjamin Braddock outside to the pool during a cocktail party. The older man drapes his arm over Braddock's shoulder, turquoise pool water glimmering in the background.

"Plastics!" says McGuire. "There's a great future in plastics. Think about it. Will you think about it?"

I *am* thinking about it. *Plasticity*—or more precisely "neuroplasticity"—how the brain changes throughout life. Psychiatric disorders clearly cause negative neuroplasticity—that is, they damage the brain, especially if untreated. And effective treatment might—just might—induce a process of positive plasticity, a kind of brain healing.

Soon after I came to NYSPI, construction began on a new MRI facility to be used for research purposes. At innumerable meetings, I began hearing how the new scanner would be shoehorned into the institute's sub-basement near the kitchen and maintenance areas, how the huge magnet would be slid through a hole cut in the side of the parking garage ramp and placed in the lead-lined scanner room. After months of planning, visits by engineers, and upgrades of hardware and software, our world-class 3T brain MRI scanner was securely anchored to Manhattan bedrock, shielded from the vibration of passing traffic and trains. Finally, just a few months ago, the magnet was powered on.

As clinical director, I've been helping make sure our patients will have safe experiences here with our new MRI machine. I'm glad to help, sure. But I keep fantasizing about doing my own studies with the MRI. In a way, it's as though a new era of psychiatry is beginning, what you might call the Age of the Scanner, since the MRI is so central to the amazingly powerful new ways psychiatry is investigating the brain. Now we're not only going to be able to measure symptoms of our psychiatric patients but actually see their brains at work, to see how our treatments change their brains' function and connectivity. It's an amazing thing, to witness the beginning of another new age.

The new MRI director at PI is Bradley Peterson MD, a young researcher from Minnesota who trained at Yale in both psychiatry and neuroradiology. One morning after our departmental grand rounds I run into him at the elevator and ask whether it might be possible to talk about working together, perhaps to combine MRI imaging with one of my clinical trials with a new antidepressant medicine.

Brad is tall, thoughtful, deliberate, friendly. A week later, we meet at his office on the second floor of the Pardes Building. Brad shows me brightly colored MRI images from one of his recent studies. By measuring "cortical thickness," he has found that people with major depression have thinning of the outer layer of the frontal and temporal cortex; that is, they have less gray matter "volume" than people without depression. "Moreover," Brad says, "people at high risk for developing major depression, but who have never been depressed, *already* have cortical thinning."

"So is cortical thinning a risk factor for depression or a result of depression?"

"Good question. Maybe both."

"What would be the effect of treating with antidepressants?" I ask. "Would it make their cortexes thicker?"

"Also a good question," he says. "We don't know."

I realize something, as Brad talks: I am pretty good at doing clinical trials, yet I still know next to nothing about neuroimaging. On the other hand, from what Brad says, the neuroimaging experts don't know much about the complicated and often frustrating business of running clinical trials. These are two entirely different skill sets. It is possible that, by combining forces, we could do some cool studies.

A few weeks later, I pitch my idea to him. This type of study is rarely done, he tells me. Most MRI studies of depression are relatively small, scanning perhaps a dozen people. Very few compare medicine treatment with placebo. And even fewer repeat the MRI scans over time.

But it will be possible to do? I ask.

"Sure," Brad says. "Let's go for it."

We begin to sketch out a collaborative project, combining a double-blind placebo-controlled antidepressant medication study with repeated MRI scans. We decide to do MRI scans before our patients start medicine (or placebo) and scan them again after a few months of treatment. Brad has also been collecting brain MRIs of people *without* depression, so we can compare our depressed patients to them, matching them for age, education, and sex.

Since we are going to the trouble of getting people in the scanner, Brad advises that we do as many types of scanning sequences as possible. Anatomical scanning, of course, to look at the way depression and its treatment affect the structure of the brain. But also other types of scanning: diffusion tensor imaging, or DTI, which evaluates the nerve tracts connecting different parts of the brain. Magnetic spectroscopic imaging, MRSI, which will allow us to measure the amount of different brain chemicals in various regions of the brain. Functional brain imaging tests, fMRI, where we can ask our subjects to look at faces with different emotional expressions, to see what parts of their brains are activated by seeing happy or sad faces, for example. "Cognitive conflict" tests (for example, showing the word "RED" written in green ink and asking the patient to tell you what color they are seeing). And resting state imaging, rs-fMRI, just asking the patient to lie still in the scanner and let their mind wander, to see what parts of the brain activate when the mind is at rest.

"How much time will all this take?"

"Probably about two hours per scanning session," Brad says. "Once the subject is inside the scanner, they are generally okay to keep going."

Soon, Brad and I apply for a grant, with our double-blind placebo-controlled design, with repeated "multimodal" MRI scans before treatment and repeated after ten weeks of either medication or placebo treatment.

Several months later, much to my surprise, we get funded.

In a way I guess I have started undergoing my *own* process of appropriately adventurous exploration.

Soon, my work with patients in my private practice begins to change as well. When I was a psychiatry resident, starting in 1980, back in the age of psychoanalysis, my goal was to help patients "work through" their psychological conflicts; the means was intensive psychotherapy, several times a week. When I went to Manhattan Medical Center in 1984 and became a devotee of *DSM* psychiatry (the era of psychiatry based on making clear diagnoses based on lists of symptoms of disorders, starting with the 1980 publication of the *DSM-III*), our goal became to treat the main symptoms of each disorder, whether with a particular medication or with a targeted "evidence-based" psychotherapy, until they started to fade away. But neither psychoanalysis nor *DSM* psychiatry can do more than guess what is going on in the brain.

Now? Now, in this emerging Age of the Scanner, what is the goal of treatment?

I don't know actually or how to measure it, much less how to achieve it in this new era of neuroscience-inspired psychiatry. What treatment methods are best to use? I do want to help my patients work through their conflicts, sure. And I want their *DSM*-disorder symptoms to fade away, so I am not about to give up on evidence-based psychotherapy or medication. But increasingly, those goals seem unnecessarily limited. Why not aim for recovery?

A strange process has thus been set in motion. As I sit in my office talking with patients, I begin to see through their skulls, into their brains. Not literally, of course, I'm not psychotic (I hope!)—but I start to imagine different brain centers at work. So, when a young college student describes his incessant, terrifying panic attacks, I can visualize his amygdala overreacting and the dulled response of his anterior cingulate cortex, which usually regulates fear signals emanating from lower brain centers but is now flooded, overwhelmed, and then the gushing secretions from his hypothalamus to his pituitary to his adrenal glands, and on to his autonomic nerve pathways, causing profuse sweating, heart pounding, and a sense of suffocation even as he faces me, telling his story.

This is really strange. Beyond that, beyond looking under the hood of the brain, so to speak, I begin to wonder: How might we change the abnormal hyperactivity (or hypoactivity) of these brain centers? If disorder causes negative brain changes, is there a way to stop the damage and even to start a process of recovery?

I know I am on thin ice here, going far beyond that data of the early 2000s. But it gives me pause. Getting symptoms under control has been essential with Hank and Aline—their improvement starting only months after symptoms faded away—but that seems like only the beginning.

Shouldn't you try to help the patient have as healthy a brain as possible? New MRI studies that Brad showed me demonstrate that depression causes shrinkage of parts of the brain, similar to the changes in the blood vessels in high blood pressure and the pancreas in diabetes and other organs in a host of other illnesses.

Maybe there is a way to improve brain health. Aline seems like a perfect example, though I obviously have no "proof' of her brain changes. But her profound fear reactions have faded away after half a decade of intensive meditation and yoga and breathing exercises, helped along by better nutrition and avoidance of alcohol, plus regular psychotherapy. Maybe Aline has been able to train her amygdala to become less hyperactive, maybe—if the brain's safety center really exists—maybe she has been able to nudge it back toward normal. And, after her terrors have faded, Aline has become able to love again: which brain circuits, I wonder, are involved in *that* process?

I am in continual ferment. I realize that several patients in my practice have begun jogging or swimming regularly and have improved their diets, with anything from vegetarianism to Mediterranean to vegan diets and various micronutrient supplements, and subsequently their depression or anxiety disorders have improved far beyond what medication or therapy accomplished. As I keep reading, I find studies showing that exercise increases the brain's "neurotrophic factors," hormones that enhance brain cell growth, causing them to become larger in size and more profusely connected to other brain cells. And possibly even causing the birth of *new* brain cells, or "neurogenesis," which was previously thought to occur only in childhood or adolescence.

Studies of meditation and mindfulness are now starting to appear, I find, demonstrating that meditation mindfulness training can decrease hyperactivity of brain centers related to fear and anxiety. And such approaches—exercise, diet, mindfulness, meditation—can increase the length of "telomeres," cellular makers of biological aging, can lead to more "hormetic," or healthful, stress, and can build stress resilience. Even psychoanalysis seems oddly relevant now: maybe the process of talking day after day to a nonjudgmental therapist, revealing whatever comes into your mind, allows for activation of the brain's safety center and causes "extinction" of fear reactions, as well as enabling better "top-down" control from the executive parts of the brain.

Clearly, I don't have it all sorted out. Every week, every month, I come up with new ideas in my practice and abandon others. But I am hardly alone. In my discussions with other New York State Psychiatric Institute doctors, my Columbia colleagues, and other psychiatrists around the country, at conferences and online, I discover that I am part of a movement, one of thousands of psychiatrists and psychologists and neuroscientists, and that we are *all* in a ferment, beginning to wrestle with these discoveries and trying to come up with ways to make connections across a what has always been a vast abyss between the lab and the clinic. Not only have many of us added MRI scanning, PET scans, or a host of other brain imaging technologies to our studies, I realize, but in a sense we have all *become* "scanners"—we are scanning the scientific literature, and we are gazing at and listening to our patients anew. New hypotheses keep bubbling up, hot to be explored, investigated, perhaps discarded. It is OK to be wrong in this strangely exciting new world. One thing for sure, though, once you start thinking about these things, there is no way to stop: the world has irrevocably changed.

SAFETY LEARNING NOW

As of 2022, where are we? For a number of years, Dr. Rogan's work on safety conditioning was continued by another young researcher at the New York State Psychiatric Institute, Dr. Daniela Pollak. Dr. Pollak completed a study showing how safety conditioning could be successfully translated from a mouse model to humans. She also has done a study suggesting that both behavior therapy *and* medication treatment can help the brain's safety system to recover. These two treatment approaches work in different ways, yet somehow both medicine and behavioral treatment can enhance the brain's ability to transmit signals that the world is again safe. She even writes a paper suggesting that safety conditioning can be explored as an antidepressant treatment in itself.

I emailed Dr. Pollak, who since then has moved to the University of Vienna. *Where does this work stand now?* Alas, she responds, things slowed down after one of her collaborators left Columbia just as they were about to extend it further in humans. After her move to Vienna, it has taken a while to set up her lab there. She is now looking for collaborators. Would I be interested?

The safety-center work has been extended in fascinating ways. Tanja Jovanovic, a researcher at Emory University in Georgia, has been studying safety learning both in animals and humans. Her work shows that abnormal safety conditioning may be a crucial part of post-traumatic stress disorder (PTSD). People with PTSD have intensified fear responses, which may be set off by TV or movie scenes, by loud noises, by nightmares and memories. They have flashbacks and startle reactions in response to these "cues," as if each time they are experiencing the trauma anew. In essence, they are unable to forget: their terror just doesn't fade away. Something in their brains prevents these responses from "extinguishing" over time. But why? Why don't they forget? Jovanovic's work suggests what the puzzle's missing piece may be.

PTSD may consist not only of impaired "fear extinction" but *also* of impaired "safety learning," an inability to "discriminate" a safe environment from a dangerous one. If you can't learn to *feel* safe, no wonder your fears don't fade away! Thus, not only may people with PTSD be unable to forget their fears—their brains may also be unable to "learn" when the world is actually safe for them again. Many people experience traumas but *don't* develop PTSD; maybe their brains *can* still learn when things are safe, whereas for some reason other people can't relearn safety. Given the many thousands of Iraq and Afghanistan war veterans, commonly suffering from PTSD, who return to the United States and find their lives in ruins, and an astronomical suicide rate among vets, there's an urgent need for understanding what blocks safety learning in those people and for developing more effective treatments.

Could treatment specifically focused on enhancing "safety learning" improve the outcome of PTSD? A 2017 study by Vasiliki Michopoulos and coworkers at Emory University suggests this may be the case. "Dysregulation" of the hypothalamus-pituitary-adrenal system—with increased production of stress hormones like cortisol—is thought to cause the symptoms of PTSD. In work that grew out of "translational rodent studies," they treated war veterans with PTSD with dexamethasone, a steroid medicine that blocks the body's own production of cortisol. The vets showed increased "fear extinction" *and* better "safety discrimination" after treatment.

For years, I have been frustrated in my work with PTSD patients, hoping to see someone replace intuition with data. And for years, I have only seen bits and pieces of what clearly is a grueling process of scientific exploration. "But there's a study to do there, *someone* should do it," I keep thinking. I keep scanning scientific journals, hoping to see it published. I keep hoping to convince one of my Columbia colleagues who is an expert in PTSD to do it, looking for collaboration.

Now, finding it, I can't help being amazed: *So here it is*—a bridge from lab research to useful medication treatments in the clinic! But it is only one step: what about adding "safety conditioning" to *behavioral* therapy treatments? Is safety learning also impaired in agoraphobia and panic disorder? And, for that matter, is it impaired in OCD and depression, too? What about doing MRI and PET studies to trace the circuits of safety learning as people respond to treatment? This is just a beginning!

But back to my patients.

Three years after their joyful jaunt to Mexico, Hank and Josie pick up their kids and move to the Pacific Northwest, lured by job offers they can't refuse. Parenthetically, you can say that Hank feels safe enough to leave the all-too-familiar East Coast for a new life in Oregon.

Since then, I've treated dozens more such patients who have gone from states of continuous fear and avoidance to a new sense of confidence and safety.

I think in particular of two young women: one with severe OCD and social anxiety disorder, the other with unrelenting, suicidal chronic depression. Both were college dropouts, both were unemployed, living with their parents, watching their classmates zoom ahead in life.

After their disorders go into remission with medication and courses of behavioral psychotherapy, both gradually return to their lives. Each of them, finding the world "safe" again, finishes college (in one case, half a dozen years after her high school classmates), each enters graduate school, each gets her driver's license and moves into her own apartment. Each starts dating.

They are entirely unaware of one another's existence, of course, or of their eerily parallel trajectories.

And Hank? Hank keeps in touch. He emails me holiday photos every December, showing the family in exotic locations, the four of them wearing bright holiday sweaters on ski slopes or tropical beaches. They wear gaudy Christmas sweaters, and they smile like crazy for the camera.

CHAPTER 11

Curing Families

Genes, Circuits, and the Frontiers of Treatment, 2005–2009

For millennia, doctors of all specialties have dreamed of curing families. In hospitals and clinics, in offices and operating rooms, we are all-too-keenly aware of devastating illnesses that pass from one generation to the next. Some are infectious, like syphilis; others, like juvenile-onset diabetes, are genetic. Still others, like alcoholism, may be a dance of genes and behavior—but they too can pass from parent to child to grandchild, a strange unwanted heirloom. Regardless of origin, though, family illnesses may wreak havoc for centuries.

In confronting such illnesses, we doctors invariably yearn to improve the fate of an entire family. Make the right diagnosis, prescribe the proper treatment, and you might change the life not only of the patient who sits before you but many others as well, some as yet unborn. Such urges certainly predate any ability actually to *cure* disease—they are inherent to the healing professions. In recent decades, though, with unparalleled medical advances such as new drugs and surgeries and the beginning of genetic engineering, doctors' age-old dreams seem to be finally coming true, especially with the new technologies I'm seeing in my work at the New York State Psychiatric Institute: the new research on brain functioning, on the effects of genes on behavior, and the effects of life experiences on the functioning of our genes. Our new era, which I'm beginning to think of as the Age of the Scanner, may offer hope.

In psychiatry, which for so many centuries could barely dream of curing a single patient, much less an entire family, such advances may have begun to arrive.

Such is the case for a patient I will call Maureen, whom I first see for a consultation in the fall of 2005, several years after my move to the NYSPI. In psychiatry, though my colleagues have long been aware of the urge to cure families, we have rarely been able to follow through with what our cardiologist or infectious disease specialist cousins routinely achieve—that is, definitive diagnosis and treatment. In fact, during psychiatric training, especially during the decades that psychoanalysis reigned, we are trained to guard against such grand hopes, such "rescue fantasies."

The psychopharmacological revolution, the introduction of effective new medications for psychological illnesses, began to change this, especially in the early 1990s, when we were flooded with new SSRI drugs that often worked better than older medications such as the tricyclics and MAO inhibitors, and even if they were not more effective, at least they were easier to tolerate—and far safer. The psychotherapy revolution also has contributed to psychiatry's increased potential to cure, for example through the introduction of evidence-based therapies like cognitive behavioral therapy for treating *DSM* disorders such as major depression or panic disorder.

Maureen's story is therefore not unusual. She tells me she is thirty-three years old, married and living in New Jersey, and that she suffers from paralyzing anxiety that pervades almost every moment of her waking life. She arrives at my private office on the Upper West Side of Manhattan, on a leafy side street off Central Park West, accompanied by an older woman, who I assume is her mother and who stays behind in the waiting room. A pale woman with freckles sprinkled from her forehead down across her shoulders and arms, Maureen stands tall and athletic, with long wavy red hair, and would appear strikingly beautiful if not for her strange manner. As she takes a seat across from me in the consulting room, she seems uncomfortable in her very soul. Her eyes dart about, her breathing is shallow and rapid, and her shoulders hunch as if protecting against invisible blows.

She hurries through her story: she is married and has a twelve-year-old-daughter, Tina, and had to quit her part-time job in a local real estate management office because of what is happening. How unbearable her life has been in the past two years, ever since she suffered a massive panic attack while driving on the Jersey Turnpike, an explosion of suffocation and terror that led her to veer across several lanes of traffic and stop abruptly in the median strip, convinced she was dying from a stroke. "It was a miracle I didn't crash!" For over an hour, she had sat in her car sobbing, until finally she summoned the courage to pull back into traffic.

Since then, Maureen's life has been one moment of terror after another. Barely able to eat or sleep, she has been in and out of emergency rooms, and her husband is at his wits' end.

Even in the room—breathless, panting, flinching, and flushing—Maureen seems on the edge of catastrophe.

"Are you OK?" I ask, and she bursts into tears. She's doomed, her life is over. For months, her family doctor has given prescriptions of Valium and Elavil, the first of which only briefly calm her terrors, the second of which leaves her groggy and disoriented. A year ago, she took a leave from her job. Now she can barely leave home.

As I listen to Maureen, it becomes clear that she "meets criteria" for the *DSM* diagnosis of panic disorder. The recent outbursts of terror, initially several times a day and now almost continual, make her life impossible. They have extended into agoraphobia, which is also common and, more recently and not surprisingly, into major depression. But her symptoms aren't entirely new: Maureen dropped out of college in the middle of junior year because of panic attacks, and since that time she has quit several jobs because of anxiety triggered by traveling away from home. Disabling rituals and compulsions, perhaps developed as a way to manage her anxiety, consume several hours of every day.

For some reason, her condition has continued to worsen, to the point that it is becoming increasingly difficult for her to be home alone. At times she even keeps Tina home from school just to have some company, which she knows is despicable. She hates herself for that. Even telling me about it, weeping, gasping, I see, makes her panic rise.

First, I need to help Maureen calm down, in the room, before we can even get her full history. I point out how even now she is hyperventilating—does she notice that?

"Yes," she says, shuddering, but her tears stopping. "I'm always gasping for air, I can never catch my breath."

"And that may be what sets off the panic attacks," I add. I explain how asking patients to hyperventilate was often used as a trigger in the lab to cause panic attacks or as a way to help people in treatment.

"Why would anyone *want* to set off a panic attack?"

"To help you gain control of your anxiety," I say, "to realize that it's not just something that happens to you. Anyway, we're not going to do that today."

Over the next few minutes I instruct Maureen in how to do diaphragmatic breathing, deep slow breaths at a slow rate of ten per minute, and, holding one hand against my solar plexus, I practice deep breathing with her.

"Do you feel any difference?" I ask after several minutes.

"Yes," she says, "I am calmer."

"You do look calmer." I tell her she needs to practice this regularly, at least twice a day, and that there are also muscle relaxation exercises that can be combined with the breathing exercises and tapes she can use to help control her surges of anxiety.

Maureen has brought a huge folder from her various doctor and ER visits—reports of recent physical exams, lab tests, EKGs, an echocardiogram—all of which, as I flip through the pages, appear normal. We can be reasonably confident that she doesn't have some undiagnosed medical illness and that, despite her fears, she is quite unlikely to drop dead.

My conclusion, after getting her history: In addition to a *DSM-IV* diagnosis of panic disorder, Maureen also has agoraphobia and possibly obsessive-compulsive disorder. Her diagnoses aren't unusual, and the treatments I prescribe that day are not unusual either as of the late 1990s. I am fairly optimistic that they will help.

Before the session ends, I write a prescription for Zoloft, one of the selective serotonin reuptake inhibitor (SSRI) medications. I also recommend that she start psychotherapy—not the kind of open-ended psychoanalytic therapy I had been through with Dr. Veltrin decades earlier, but cognitive behavioral therapy, CBT. CBT—in which I trained during my first years at MMC—can help her deal more effectively with anxiety by helping her "restructure" automatic thoughts and behavioral reactions. Beside the deep-breathing exercises we did today, it also includes a process of changing habits to try to face fears and stop "avoidance behaviors." All of these approaches are "evidence based"—meaning that studies have shown they are effective in treating people with these disorders.

Maureen has a tough time starting the Zoloft. Initially, it causes her to feel *more* anxious, so she needs to begin with a low dose and only gradually increase it. She practices the diaphragmatic breathing exercises and keeps a record, per my suggestions, charting her anxiety level between zero and one hundred, how it changes from before the exercise to afterward. Reliably, after just a few minutes, her anxiety will drop from the eighties or nineties down to the fifties or sixties, that is, from extreme to moderate. That simple exercise helps her realize that she is at least *partly* able to control her level of anxiety; that's not optimal, but it's better than nothing. After a bit of searching, she finds a trained CBT therapist who practices near her home in New Jersey and begins seeing her on a weekly basis. CBT techniques—progressive exposure, response prevention—gradually help Maureen take more risks on a daily basis, to confront her fears about being alone, her terror of leaving home. Soon she begins driving again and stops keeping Tina home from school.

Over time, Maureen's cumulative improvement is profound. As panic and obsessions start melting away, her life opens up. For the first time since childhood she can travel without fear. She returns to work at the real estate firm and researches how to reenroll to complete her BA. Her panic attacks go from daily to almost never. She feels protected by the medicine but even more so by her breathing and relaxation techniques. She says she knows how to cope with her panic now.

Then the real work begins. Maureen asks: Will I agree to see her husband, Tom? He too suffers from severe anxiety. The events of September 11, 2001, when he was working in downtown Manhattan a few blocks from the World Trade Center, set off a chain of frightening events, sensitizing him to all kinds of risks. Going to job sites is becoming a real problem for him, which has led to him often turning around and going home. Recently his boss wrote him up.

Tom cancels two appointments before finally making it to my office. A lanky, sheepish guy, he admits to being appalled by needing to see a shrink. He keeps calling me "Doc." "Doc, I can't go on the subway. I can't go in high buildings, which is a problem since our office moved to the thirty-first floor. I have panic attacks too, Doc, just like Maureen. Worse in fact. I had a drinking problem for a while because that was the only thing that helped. Doc, what's wrong with me?"

His attacks include classic symptoms: severe palpitations, a sense of suffocation, feelings that he is about to die, tingling in his fingers and toes, ringing in his ears, a sense of floating away from his body, even paranoid ideas at times, a fairly unusual symptom. He is continually terrified of having another panic attack, especially at work, and of being humiliated in front of coworkers. Whenever he is assigned to a new job site, he'll be up half the night, planning each step of tomorrow. While his symptoms were generally tolerable when his company was mostly reconstructing storefronts and building low-rise housing, now they are constructing taller buildings, and it is all he can do to steel himself to get into an open elevator, emerging on windy floors with no walls.

"Doc, it's kind of funny, me and Maureen, both with panic, d'you see a lot of couples with that?"

Indeed, their shared suffering is one of the things that brought them together to begin with. They met in seventh grade, he tells me, both refusing to take a field trip from Jersey into Manhattan, and locked eyes while sitting on a hard wooden bench outside the principal's office, awaiting detention. They have a solidarity of disorder: both hate bridges, tunnels, heights, blood draws, parties and other social events, and open spaces. Their shared suffering has forged an intense intimacy. They started at a local college as friends and began to date junior year, just before she dropped out of classes, and they married right after he graduated. A Darwinian would call it "assortative mating," a pattern in which individuals with similar genes or other characteristics pair up more frequently than would be expected under a random mating pattern.

Tom's panic attacks have worsened in recent months, too, but he has been so consumed with his worries about Maureen that he wasn't able attend to himself until she started to improve. In an effort to control symptoms, he now drinks up to two six-packs of beer a day.

I begin working with Tom. First priority, the alcohol. If he can't manage to cut back on his own, he'll have to go to AA. He is okay with that but is reluctant to consider CBT or any kind of therapy, saying it is too expensive.

By our next meeting the following month he has stopped drinking almost entirely; he says he's ready for pills. "Let's see what the Prozac does, Doc." A regular starting dose, 20 milligrams per day, has a modest effect; then I increase it over a few months to a full dose, 80 milligrams.

If anything, Tom has an even more pronounced response than Maureen. His panic attacks completely disappear. There are plenty of tests on a daily basis in Tom's life. Traveling to work sites. Running meetings with contractors and clients. Attending work conferences, going up the open-sided elevator in new construction sites, sitting in godawful traffic, even going to his company's annual Christmas party on the thirty-first floor in an office with a vertiginous view over the Palisades and the Hudson River. In dramatic contrast to past experiences, now Tom feels amazingly calm, pacific; nothing can rile him. His metamorphosis is wonderful in so many ways—he actually starts having fun hanging out with his coworkers, making new friends, joining a rec basketball league. It's wonderful but also disorienting.

"I kind of got used to having anxiety all the time, Doc," he confesses. "There were so many things I couldn't do. Now I just basically feel OK, I can do pretty much whatever I want. Which is good, but it's kind of weird too. No limits, you know?"

Next comes Maureen's mother, Suzanne, who accompanied her daughter to the first visit. For all Maureen's life, Susanne has been her rock and her nemesis. Nagging, insecure, fearful, clinging, manipulative, but "always there for me"—Maureen has no shortage of adjectives for Susanne. They have been each other's confidantes, comforters, life preservers. I am both eager to talk with Susanne and a bit apprehensive.

On the day of her mother's first visit, Maureen returns the favor, accompanying her fearful mother. Maureen drops her mother off at my Central Park West office, then heads out to get some coffee at the Starbucks that just opened on the corner of Columbus Ave.

Susanne has a worn, gaunt appearance and appears older at least by a decade than her sixty-five years, as if consuming worries have led to premature aging. She is perceptive and a bit sharp-tongued. "Maureen and Tom are so much better," she tells me. "Their dynamic is completely changed. Now you're going to fix me!"

Long ago, Susanne tells me, she trained as a classical pianist. Despite her near-virtuoso technique, she was unable to give concerts or recitals because of disabling anxiety. For many years she was barely able to leave home because of severe panic. She fell back on teaching piano to kids. It was so hard even traveling to her students' homes that she gave lessons at her home. She became the family's breadwinner after separating from her husband, an alcoholic, when Maureen was seven years old. He too was an anxious man. A World War II infantryman, a steamfitter, he drank to soothe his anxiety and developed a severe alcohol dependency, but after he committed to sobriety and joined AA, they got back together.

Susanne meets criteria for what the *DSM-IV-TR* (the most recent *DSM* of that time, published in 2000), bringing two diagnoses together, has begun to call "Panic Disorder with Agoraphobia": "anxiety about being in places or situations in which escape might be difficult." It's crystal clear that day, in evaluating Susanne, how a life-long, untreated psychiatric condition has not only led to a mostly miserable life but also has poisoned her mother-daughter relationship. Continually fearing the next panic attack, especially after her husband left, Susanne grasped unceasingly at Maureen, keeping her home from school as a child (as Maureen began to do with Tina), interfering with friends and boyfriends during her teenage years, basically driving her batty.

So, what to do for treatment? Susanne pooh-poohs any idea of therapy—she never improved over two decades of psychoanalysis. As far as she is concerned, therapists are useless; she only wants pills. "Give me what you gave Maureen."

Suffice it to say, Susanne responds beautifully to Zoloft as well. All she needs is 50 mg/day, first by my prescription, then continued by her general practitioner. The way I know is that Maureen's dad calls to thank me—for the first time he is able to enjoy his retirement.

Susanne soon begins taking the PATH train into the city nearly every week, going to the latest exhibits at the Met, at MOMA. Even dragging her husband to spend weekends at hotels, so they can attend concerts and Broadway musicals without having to rush home at night.

This is not the end. What becomes clear is that Maureen's entire "pedigree" is stricken with severe anxiety disorders. Nearly all the women in Maureen's family have suffered from intense panic attacks or incapacitating depression; the men either have panic, alcoholism, or both. What really drives this home to me is the day that Maureen mentions that her grandmother, Moira, a ninety-two-year-old woman, has *also* suffered from crippling anxiety her whole life!

No need to worry that Maureen will be pressing me to see her too—luckily, Grandma Moira lives in London and is reluctant to travel. Maureen shows me pictures of her grandmother: a wizened, bright-eyed, gray-haired woman standing in her garden in a village north of London. An amateur botanist, Moira is passionate about native British plants; being housebound, she has cultivated a magnificent garden over decades. Hearing of her daughter and granddaughter's response to medication, Moira too decides to try an SSRI. To make a long story short, she achieves a remarkable response. Prescribed Cipramil by her GP in the National Health Service (that being the British brand name for citalopram, or what Americans know as Celexa), she is liberated from nearly a century of severe anxiety—and she soon declares to her family (communicated via Maureen) that she doesn't care how old she is, she is going to enjoy life to the fullest. If she can travel without terrifying anxiety, if she can visit museums and shops for the first time in decades without attacks, she is heading into central London, and nothing will hold her back!

All in all, it is like a Russian folktale, in which a traveler comes to a hut in the midst of a dark forest and is introduced to one generation after the next, each tinier and more wrinkled than the last. Only in this case, I was uncovering a medical saga and getting a lesson in how disorders "breed true."

Then, perhaps inevitably, arises the question: will I see Maureen's twelve-year-old daughter, Tina? Maureen preempts any discussion by bringing Tina to my office one day, unannounced, on their way to performance at the ballet. Can I just talk to her for a few minutes? She is a string bean of a girl, a passionate dancer, and Maureen waves her into the other chair in my office.

"Tina, tell the doctor, you're having anxiety too, right?"

Tina flushes. She tells me how she had a very bad attack at Newark Airport on her way to dance camp this past July, so bad that she had to be taken off the plane. "And I'm always worrying," Tina adds. "A lot of times I get so upset that I throw up before school in the morning."

"It's true," Maureen says. "There's no reason for it, she loves school, she has a ton of friends."

"But I get so worried!"

After a few minutes, Maureen sends Tina out to the waiting room, so we can have our session. She closes the door. "Doctor, do you think you can give Tina medicine too?"

I decline, not being a child psychiatrist, but give her some names of people who could provide a consultation. You can guess the rest of Tina's story: she too responds to an SSRI.

To celebrate, the whole family picks up and heads to London to visit Grandma Moira. Susanne and Chuck and Maureen and Tom and Tina stay in a hotel for ten days, visiting Moira's small garden home an hour north of the city, and they have a blast at the British Museum, the Tate, and the Royal Shakespeare Company. It has been thirty years since Susanne saw her mother, who hadn't come to Maureen and Tom's wedding and who has never met Tina.

After all this, I am beginning to feel like a shill for the pharmaceutical industry. Normally I am rather conservative in my practice. I make every effort to recommend therapy before medicine and avoid claiming remarkable results for any treatment. But in the case of Maureen's family, I must concede that there is a profound response to a single class of medication, one that blocks reuptake of a particular neurotransmitter at a specific type of receptor.

Perhaps, I hypothesize, Maureen's family has some kind of genetic disorder, possibly a defect in their brains' serotonin transporter system, which predisposes them to severe anxiety. After all, in a fascinating new paper in *Science*, published in 2003, Avshalom Caspi and colleagues have just reported that individuals with one or two short arms (the so-called s alleles) of the *5-HTT* serotonin transporter gene (the gene for the receptor in the brain that inactivates serotonin) get depressed more often after stressful events than individuals with two long arms of this gene. This could explain why depression and related conditions often run in families. Maureen's entire family might have that genotype, I muse, or perhaps a related one that codes for anxiety instead of depression.

I begin to feel pretty good about things: not only my diagnostic acumen, which connects so neatly to the latest findings in neuroscience, but also my ability to manage a family's illness, which meets all the usual criteria for dreadfulness. In a way, this is emblematic of my transition from the trenches of Manhattan Medical Center to a new life at NYSPI. Whereas at MMC we were always worried about the next financial crisis, the next round of cutbacks and layoffs, and how to accommodate the throngs of desperate patients that surged through our doors that we barely had time to think, here at NYSPI, or PI, as we call the institute, we have time to think about things, to read the latest research, and to make connections between what we see in the clinic and what researchers are doing in their labs, studying isolated brain cells or "animal models" of psychiatric illnesses. We can cultivate a new flexibility of mind, going between one setting and another, but also trying different models on for size, different explanations for what we are seeing in patients we encounter in our daily practice.

Now, attending conferences and meetings at PI, in discussions with my new colleagues and students, I am becoming acutely aware how new research can make us look anew at routine clinical encounters. Whereas in Age of the Couch, Maureen's mother was treated (unsuccessfully) for years with psychoanalysis, in the *DSM* era, the Age of the Clinic, we were able to come up with solid diagnoses of her entire family *and* apply effective treatments that changed their lives. There were many cases in my practice similar to Maureen and her family in which decades-long patterns of suffering and dysfunction began to give way to new life opportunities and satisfaction. In fact, transformation wasn't such a surprise anymore.

But there is something else: in this new, emerging age, the Age of the Scanner, which I joined when I came to the New York State Psychiatric Institute. Now we are starting to see *beyond* the *DSM-IV*. To see through our patients' skulls into the workings of their brains, so to speak, to incorporate our new discoveries from a host of scientific disciplines into the care of patients.

It is exhilarating to try to understand their strengths and their suffering through our new neurobiological lenses, rather than the (far more comfortable and familiar) perspectives of psychodynamic and *DSM*-based psychiatry. Exhilarating and almost magical.

For instance, in my work with Maureen, I begin to think about brain circuits, and about genes, and even about what are called "epigenetic" changes. Several lines of new neuroscience research converge on a fascinating conclusion: Certain brain circuits are overly active in a number of psychiatric disorders—fear circuits in post-traumatic stress disorders, for instance. Certain gene types are thought to increase a person's risk of developing bipolar disorder, depression, and schizophrenia. And *epigenetic changes*—modifications of genes that occur *during* life—can take place especially as a result of highly stressful events or exposure to trauma. These experiences can have long-lasting, even life-long effects. Extreme life events can change the actual structure of the gene, by a chemical process of methylation or acetylation—the binding of carbon chains to a person's DNA, either turning the genes off or on, potentially for the rest of that person's life!

Taking Maureen as an example: her family seems to have some "genetic risk" for high anxiety, which has probably caused elevated activity of brain circuits from early in life. A child may have a calm temperament or a fearful one. Such genes and circuits may increase risk but don't necessarily lead to disorder. Life events, such as trauma, can trip the switch. The story in Maureen's family is that Grandma Moira's immobilizing anxiety began when she was a child in London during World War II and spent nights in terror, hiding from German bombs in her building's cellar. The family has been anxious, as they see it, ever since.

Perhaps, I wonder, an epigenetic change from the Blitz bombings triggered many generations of suffering. Does this sound farfetched? Studies of mice do show that trauma can be transmitted from one generation to the next and to following generations as well. Each generation, having lives dominated by anxiety, has then raised their children in a household dominated by fear. New research has shown that anxious (or depressed) parents can produce environments in which their kids have a higher risk of developing anxiety—at least partly because of epigenetic changes brought on by living in high-stress environments. *Anxiety can be passed down through the generations.* A child at genetic high risk of developing anxiety, if brought up in a *low*-anxiety household, may never develop panic attacks. The right environment may actually protect them against their anxiety-causing genes. Studies of mice have shown that the offspring of traumatized animals—even if raised by untraumatized "dams"—bear the traces of trauma in their bodies and brains, the results of "epigenetic" changes—methylation or acetylation of DNA—that are transmitted from one generation to the next. And not just children of the traumatized mice but mouse grandchildren and great-grandchildren as well. Which at least partly explains why many-generation effects of the Holocaust in families of survivors, the depression, anxiety disorders, and other problems that we saw in families we treated at MMC—grandma's depression, Dad's alcoholism, grandson's heroin dependence and suicide attempts—often continue to afflict them over half a century later.

So not only am I excited about the changes in Maureen's family, but their case seems to me emblematic of the new era of psychiatry now beginning to emerge, an era in which we can start to make interventions beyond the individual patient and have an impact on entire family systems, based on their shared genes and life experiences.

One day in early 2009, I get a message from Maureen on my answering machine. "I have to see you, it's an emergency."

The moment she comes through the door, she blurts out, "Tom's leaving me!"

I am speechless. "Why?" I ask finally.

Her eyes swollen, her face flushed, she slumps in her chair. "There's *no* reason. He says he just doesn't love me. He wants to get on with his life. That's what he says, everything in our lives doesn't matter, he wants to be on his own. He says the only reason we were ever together is the panic and now we don't have it so we don't have anything in common." After the sadness and despair come waves of anger and bitterness. Rage, hatred: how can he do this to them?

I am stunned. It makes no sense: their lives are so much better. It has to be some sort of mistake, an overreaction of some kind; is Tom maybe having a hypomanic reaction in the context of his life crisis? Otherwise, it doesn't make any sense to me.

"I thought he was drinking again," Maureen says. "But that's not it, and he doesn't have a girlfriend, at least that I know of. I asked him, I looked at his email and cell phone texts—I don't see any other person there . . . he could be lying, but . . . but, why—why is he doing this?"

I have no answer for her.

Tom is adamant when I meet with him the following week. It is not up for discussion, it's a done deal. He looks different. Not only a new haircut, close-cropped on the sides and full on top, a bit of gel, with his new clothes and his skinny-leg jeans, his shirt with a clipped collar, kind of a 1970s Travolta look. He has lost weight too, the beer gut he had before, the result, he says, of working out daily and swearing off alcohol. But beyond that, he has a new confidence, a swagger. "See, Doc, you've got to understand, I met Maureen when we were thirteen. We never dated anyone else. Together since we were kids, you know? We understood each other, totally, nobody else did. Sure, we had that in common, but not much else. Sorry, that's the truth. It kills me to say it. It kills Maureen too, but I don't want to keep pretending, I just can't do it."

He doesn't have a girlfriend—that has nothing to do with his decision. "Not yet," he adds. He is in talks about joining a friend's consulting company, so he can work and live in the city. Liberated from his disorder, no longer fearful of panic attacks, Tom has come to a cruel realization: he and Maureen clung to each other out of fear rather than love. What seemed to be profound commonalities, in his raw new view, was simply a shared disorder. Now they are both better, and he wants to get on with his life. "I do love Tina, no question. I'll always be there for her, I'm her dad. But that doesn't save me and Maureen, no way."

Soon afterward, he leaves them. It is awkward, in that I am continuing to treat both Maureen and Tom, and they aren't together anymore. They finally do agree to see a couple's therapist, but mostly it's for show: his mind is made up. He gets an apartment in Manhattan and begins living the life of a twenty-five-year-old.

Now, years later, I am still trying to make sense of it all.

On the research level, I have come to realize how much more complicated the Caspi story actually is. After more than a decade, the "single gene theory" for depression was shown not to hold up. The serotonin transporter gene, the short s allele, doesn't seem to code directly for depression or anxiety. Instead, the latest research shows something different and perhaps more interesting: rather than being a "bad gene," as earlier studies suggested, the short s allele may be a *plasticity* gene, making people more sensitive to their childhood environments, good *or* bad. That gene type may make people more vulnerable to stressful environments but also more *resilient* if they grow up in supportive environments. That's what fascinates me about research: some theories are supported by later studies, others are disproven, but many others show complexity and lead to fascinating new theories, ones needing further verification and testing. For Maureen's and Tom's families, high stress early in life may have led to more vulnerability, but if Tina is able to create a supportive, nonthreatening environment for her own children, maybe they will be extraordinarily resilient!

On a personal level, I keep thinking what a handsome couple they were, Maureen and Tom. I was so proud to rescue them from their disorder, which, given today's treatments, is not terribly difficult to do. But I couldn't control their lives. They were healthy again and free to make their own difficult decisions.

My success had, after all, induced a rescue fantasy: I wanted to make a happy family, but this was beyond my abilities as a doctor. All a doctor can do is treat illness. That, I realize, is a hard lesson.

CHAPTER 12

Off Label

Revisioning Drugs in the Age of Neuroscience, 1997–2023

THE WILD WEST

This is a story that spans decades. It links worlds, you might say, the Age of the Clinic—of observable symptoms in unique patients with *DSM* diagnoses—and the new twenty-first-century world of neuroscience-based psychiatry in the Age of the Scanner. And it has a personal side as well, you might say, related to my own prejudices, hopes, and intuitions about what we doctors do and don't know about the medicines we prescribe every day.

It starts in the late 1990s, with a hallway consult from one of my fellow psychiatrists at Manhattan Medical Center, Dr. Ella Farred. By this point, based on my studies and clinical experience, I have earned a reputation for knowing something about "treatment-resistant" mood disorders, whether major depression or bipolar disorder.

Dr. Farred tells me about one of her bipolar patients, Mr. W. "So, he got toxic on lithium," she tells me. "He refuses to take it again. He gained a ton of weight on Depakote, and now he refuses *that*."

Mr. W is a forty-two-year-old married man with an eight-year history of bipolar disorder, leading to three hospitalizations for manic episodes. Now he is unemployed, separated from his wife, and applying for disability.

"So now what am I supposed to do?"

"What were you thinking of?" I ask.

"Well, I could just keep him on Risperdal," she says. "But then he has a risk of getting tardive. Or . . ."

"Or what?"

"What about Neurontin?" she asks. "I hear that a lot of psychopharmacologists are writing for it off-label for mania."

I think about it. A neurological medication, Neurontin (generic name gabapentin) is FDA-approved for partial seizures in adults. It is pretty safe: no interactions with other drugs, not toxic in overdose. It is not even metabolized in the body—you clear it through your kidneys, that is, pee it out unchanged—so it is especially safe for people with liver disease or other serious illnesses. Starting the late 1980s, a number of other neurological meds have been found to be mood stabilizers, useful additions to lithium, at that time our mainstay in manic-depressive illness. Depakote for one, also Tegretol—both were originally anticonvulsants, like Neurontin. A few scientific papers have suggested that Neurontin *also* works for bipolar disorder. The studies were small case series, not placebo-controlled RCTs, but they are encouraging.

"It's probably worth a try," I say. We discuss dosing: starting at a few hundred milligrams per day, up to a max over two thousand. "Let me know how it goes."

This happens back in the late 1990s when everyone is first getting excited about all the new drugs coming into psychiatry. Some are brand-new compounds; others—like Tegretol and Depakote—are "crossovers," drugs used in other medical specialties that have seemingly remarkable effects in psychiatric disorders, being prescribed "off label." "Off label" means when a doctor prescribes a medicine for a use for which it has not gotten FDA approval—a practice both common and legal, though as we will see, one that comes with its own challenges: while MDs are free to prescribe drugs off label and to explore off-label effects, off-label *promotion or advertising* by drug companies for such uses is wrong, even criminal, as it violates the FDA approval process.

As of 1997, Neurontin looks like a poster child for new crossover drugs. It is already used widely in neurology for treatment of partial seizures as an add-on to other antiseizure medicines (partial seizures being convulsions that are limited to one part of the body, in distinction to full-body "tonic-clonic" convulsions), so we already know it is safe.

Besides that, there is its "mechanism of action": Even the name "gabapentin" is alluring—it suggests to us doctors that Neurontin might have an effect on gamma-amino-butyric acid, or GABA, a brain chemical that is widely distributed in the brain and regulates anxiety (though later it is discovered that Neurontin doesn't work through GABA at all but instead through something called "voltage-gated calcium channels"!). Drugs with different mechanisms of action are always appealing when conventional treatments fail.

Then there is the need. There are so many desperate patients—ones who can't tolerate lithium, our standard treatment for bipolar disorder since the 1970s; ones who get kidney or thyroid failure or develop severe GI symptoms, or ones who have bad reactions to Depakote or Tegretol, or ones whose depression cycles faster with the use of antidepressants. We urgently need something better! And as a psychopharmacologist, a researcher as well as a practitioner, I'm always on the lookout for new treatments that might be effective, things that might help my patients and that might be worth studying in a clinical trial.

Starting in the 1980s, my colleagues and I have used many drugs off label. Take the SSRIs: we began prescribing them for panic disorder long before they were FDA approved. We prescribed them for OCD, for premenstrual mood disorder, for generalized anxiety disorder, and for "dysthymic disorder," a form of low-grade chronic depression for which no medicines have FDA approval. We began prescribing Tegretol and Depakote for bipolar patients before either was approved by FDA. Collectively we have helped hundreds of thousands of people.

How do Ella and I get the idea that gabapentin could help in bipolar disorder, otherwise known as manic depression? Perhaps one of the ever-present drug reps mentions the small bipolar studies and hands out a reprint or two of a case series of a few dozen patients who improved (but with no placebo comparison group), or perhaps we see a copy of a research conference poster. Along with an offhand whisper that some of the better psychopharmacologists in town—the ones with fancy Park Avenue practices—swear it helps their bipolar folks.

That, plus Neurontin being so intriguing on its own merits, sways us to give it a try. Most of our off-label use has been beneficial, after all.

Ella updates me on Mr. W a few months later. In brief, things have not gone well. Mr. W seemed fine at first and was greatly relieved to be tapering off his antipsychotic medication. Feeling more energetic and less stiff and drowsy, for a few weeks he appeared to be reclaiming his healthy old self.

Then things head south. He starts feeling *too* good, his wife complains. He is on the phone to CEOs with a raft of new business plans; he is reaching out to potential financial backers; he starts buying new stereo components and working out at all hours at his health club, giving unsolicited help to other members on the weight machines. Before Ella knows it, she is getting a call from the local ER that Mr. W is in a floridly manic state, in need of involuntary hospitalization. It takes several weeks in the hospital to get Mr. W back to "euthymia"—that is, to normal mood, by which time he is on a regimen of Tegretol and Risperdal, which is far from side-effect-free but at least has a reasonable chance of working.

When the evidence finally comes in, it turns out that Mr. W's case is no fluke. The problem being: Neurontin doesn't work for bipolar disorder. Not at all. In rigorous double-blind placebo-controlled studies, Neurontin is a total bust. Doesn't work as an antimanic drug, doesn't help bipolar depression either. In one study, Neurontin actually does *worse* than placebo!

COMEUPPANCE

Soon Neurontin becomes a poster child of a different kind: in 2004, Neurontin's manufacturer, by then a division of Pfizer, Inc., is cited by the Food and Drug Administration (FDA) for various nefarious practices, including unethical marketing for uses including pain disorders, amyotrophic lateral sclerosis (ALS), attention deficit disorder (ADD), migraine headaches, and, of course, bipolar disorder. The company's medical liaisons were reportedly trained to promote off-label uses of Neurontin in sneaky ways—for instance, pushing such information even when doctors didn't ask for it. The drug company actually paid ghostwriters to create articles touting the benefits of Neurontin that were signed by experts in the field, giving the impression of high-level approval for off-label uses. The fines total $430 million.

No less of an authority than Marcia Angell, a former editor of the *New England Journal of Medicine*, blasts Neurontin in her 2004 book *The Truth About the Drug Companies and What to Do About It* for its use outside the officially FDA-approved indications: "Neurontin, for example, was initially approved only for epilepsy. But, after a slow start, it grew to become a $US2.3 billion ($3 billion today) blockbuster for its owner, Pfizer, in 2003." Not only that, but "about 80 per cent of the prescriptions were for unapproved uses—conditions like bipolar disorder, post-traumatic stress disorder, insomnia, restless legs syndrome, hot flashes, migraines and tension headaches. . . . In fact, Neurontin has become a sort of all-purpose restorative for chronic discomfort of almost any type—yet there is almost no good published evidence that it works." By extension, Dr. Angell is castigating all of us physicians who have been gullible enough to use this horrible drug. Which includes me, of course, and Dr. Farred, and thousands of other well-meaning doctors.

Things are even worse than that. A 2009 paper in the *New England Journal of Medicine* shows how bad the published papers about gabapentin actually were: negative studies were often buried, left unpublished, and among the studies that were published, authors often tweaked their results to cherry-pick the most positive outcomes.

Soon no one prescribes Neurontin for bipolar disorder anymore, or if they do, they certainly won't admit to doing so.

Except for one thing: prescriptions for Neurontin continue to soar. From 18 million per year in 2004, prescriptions rise to over 45 million per year in 2016, a 250 percent increase. And the vast majority are (no surprise here) for off-label uses. So clearly, despite the bad press, doctors have hardly stopped prescribing gabapentin. Perhaps they aren't prescribing for bipolar, but they are obviously not giving up on it.

This is weird. Usually when a drug is entirely ineffective its use eventually fades away. Even drug-company buzz, patient requests, and physician habit aren't sufficient to keep totally ineffective drugs in use. Sometimes even when decent research shows that a drug *does* work, its use trails off—because it just doesn't work well *enough*. The perfect example being buspirone, a nonaddictive antianxiety drug: although many studies show that buspirone helps anxiety, few doctors prescribe it, because patients don't find it to be very effective, especially when compared to other medications like the benzodiazepine drugs such as Ativan or Valium—or the SSRIs, which also have antianxiety effects.

But Neurontin? Like thousands of other American doctors, I still find myself writing for it. At any one time, perhaps a dozen patients in my practice are on this terrible drug. Embarrassed as I might be—outed by the eminent Dr. Angell and condemned by all manner of watchdogs—I have yet to purge it from my practice. True, I am not prescribing it with the hope of controlling manic episodes but, rather, for other purposes: to decrease anxiety and agitation, to improve sleep, to help with pain. Dr. Angell is scathing about its use for bipolar, but she also castigates these other uses. When I prescribe it for such purposes, I am doing nothing wrong: I am following in the common (and entirely legal) tradition of physicians prescribing for off-label use—but am I being medically irresponsible?

The thing is, Neurontin, prescribed off-label, sometimes *works*. It is often a good second-choice medicine—less addictive than other antianxiety medicines like Xanax and far less dangerous in overdose than pain meds like OxyContin. Slightly less effective, perhaps, but definitely less likely to cause harm. (Though Neurontin may have some addiction risk, it seems to be much less than the risk of Xanax or Oxy addiction.) Sometimes, for whatever reason, it even can be life changing.

For example, Walt, a man in his early seventies. After brain surgery and radiation treatment for a brain tumor, Walt becomes explosive, easily agitated, and highly impulsive. At times his wife flees the house in fear of his fists. Lithium, Depakote, Tegretol, Haldol, Risperdal—none of them help. Overwhelmed by his angry and threatening outbursts, his wife actually files for divorce. But now, on 2,800 mg per day of Neurontin, he becomes fine, more or less back to his normal self. Both she and Walt tell me many times that Neurontin has saved their marriage.

Then there is Hakim, a patent attorney in his sixties who has bipolar disorder and chronic insomnia. Lithium has helped control his mood swings, but he finds it impossible to fall asleep. He lies awake until three or four a.m., then falls into a profound, death-like sleep, and he has great difficulty rousing himself in the morning to get to his office. Sleeping pills like Ambien and Halcion helped temporarily in the past but required higher and higher doses, eventually leading to addiction, then detox. A host of other tranquilizers, sleep aids, and sedatives have flopped. Sleep EEG studies and all sorts of expensive workups have been a waste of money. For whatever reason, a tiny dose of Neurontin, a mere 100 mg, allows Hakim to fall asleep at night. This works for years, without any need to raise the dose.

And finally, Sue Ellen, a sculptor in her fifties, with major depression and chronic pain. She suffers from diabetes mellitus, with the complication of "diabetic peripheral neuropathy"—relentless burning and tingling in her hands and feet. Pain worsens her depression, and depression compounds her pain, making it impossible to work. Neurontin, taken at 1,200 milligrams per day, nearly erases the burning, and life without chronic pain makes it possible for Sue Ellen to get back to the studio and to recover from depression.

I also see patients with a history of alcoholism. For many of them, a bit of Neurontin helps quell their desire to drink, the restless unsettledness that leads them to pick up a glass. I keep thinking of them as having "irritable brains," with twitchy neurons whose activity is somehow quelled by this odd anticonvulsant. There are other patients too, with PTSD, with severe social anxiety, for whom Neurontin provides welcome relief. I assume that other doctors who keep prescribing this medicine feel similarly. Even if the data from double-blind studies aren't particularly convincing, a fraction of their patients find real benefit from its use.

Which, as a doctor who prides himself on practicing evidence-based medicine, makes me . . . I was going to say proud, but it is more accurate to say very uneasy. I like to pioneer new uses, but I don't like to continue them without sufficient confirming data.

Years later, I still feel guilt for potentially contributing to Mr. W's rehospitalization. But even without drug reps' nefarious practices, I probably would have suggested Neurontin to Dr. Farred—or another off-label treatment like Tegretol (which didn't get FDA approval for bipolar until 2004). We were trying to make the best of bad choices.

A NEW WILD WEST

> *Recent data suggest that ketamine, given intravenously, might be the most important breakthrough in antidepressant treatment in decades.*
>
> —Dr. Thomas Insel, director of the National Institute of Mental Health, 2014

Where do we stand today with off-label use in psychiatry?

Since my 1990s hallway chat with Dr. Farred, there have been hundreds of studies of Neurontin and its cousins. The result? Basically, a zigzag from what we expected way back when. Neurontin, a bust for bipolar, now has proven benefit for pain. In fact, it has gotten an FDA "indication"—or approval—for postherpetic neuralgia, the severe pain that commonly occurs in cases of shingles. In shingles, the herpes zoster virus infects nerves, often in the torso, leading to girdles or belts of painful blisters that pop open, forming crusting, oozing scabs. The shooting pain of shingles—the intense burning, the itching, tingling, stabbing, and hypersensitivity to the lightest touch—often constitutes the worst suffering that people have experienced in their lives. For many desperate patients, Neurontin provides welcome relief. It is frankly amazing to have a nonopiate drug that can alleviate such suffering.

Neurontin is also shown to alleviate other forms of pain—especially the "neuropathic pain" seen in spinal cord injuries, diabetes (which perhaps explains my success with Sue Ellen), and in multiple sclerosis, though it does not have FDA indication for these conditions yet. Other common uses are still largely not proven. There is "suggestive evidence" that Neurontin helps alcoholics who have stopped drinking remain sober. Benefits for anxiety disorders or insomnia are not yet proven.

The Neurontin story doesn't end there, however. I mentioned earlier that it doesn't work through the GABA system but instead by its effects on calcium channels: it interacts with a "high-affinity binding site" in brain membranes, the alpha-2-delta receptors, which have recently been identified as "an auxiliary subunit of voltage-sensitive calcium channels." This different and "*novel*" mechanism is of great interest to scientists and doctors, and of course to drug manufacturers. As well as to patients.

Today we have an entire emerging class of new "gabapentinoid" drugs that work through those exact calcium channels; they are called "alpha-2 delta ligands." The best-known one is pregabalin (brand name Lyrica). Unlike Neurontin, which was first approved for seizures, Lyrica got its initial approvals for pain control. It is FDA approved for treating "neuropathic pain associated with diabetic peripheral neuropathy" as well as for postherpetic neuralgia and partial seizures in adults.

Several other gabapentinoid drugs are now under development, including imagabalin and mirogabalin. Which is good, because neither Neurontin nor Lyrica is tremendously effective. One estimate is that only 30 percent of shingles patients respond to Neurontin, compared to 20 percent on placebo. A real difference, it seems, but a small one. It would be wonderful to have better gabapentinoid medications, perhaps to help 70 or 80 percent of shingles patients.

More importantly (and often worryingly), the off-label use of drugs in medicine is hardly disappearing. In psychiatry, over half of *all* prescriptions are for off-label uses. Whether the long-standing use of propranolol for stage fright or the newer uses of the blood pressure drug prazosin for PTSD or the Parkinson's disease drug pramipexole for bipolar depression, there is an ongoing off-label boom, or, seen a different way, an off-label epidemic. In other fields of medicine, the same is true. A 2016 *JAMA Internal Medicine* article found that 11.8 percent of *all* prescriptions were written off label. Off-label use is not without risk, though: the same article found higher rates of side effects when there is less scientific evidence for benefit.

In psychiatry, the most striking (and alarming) recent off-label phenomenon is ketamine. An anesthetic first used by veterinarians, ketamine received FDA approval for use in humans in 1970, just in time to be used for wounded American soldiers in Vietnam. Back in the United States, it quickly became a club drug, prized for its hallucinogenic and dissociative effects and popularly known as "Special K."

Fast forward to the early 2000s, when studies showed it could quickly relieve depression symptoms. Really quickly: in contrast to traditional antidepressants that can take weeks to work, an IV push of ketamine can relieve depressive symptoms in as little as four hours. Benefits of a single dose can last a week or more. Ketamine works by blocking the brain's N-methyl-D-aspartate receptors, otherwise known as NMDA receptors. These receptors are widely distributed in the brain and bind to glutamate, an excitatory neurotransmitter implicated in memory functions and the plasticity of synapses. By binding to NMDA receptors, ketamine then leads to the activation of AMPA receptors and sets off a variety of effects in the brain, including an increase in neuroplasticity, a development of new brain circuits, and, of course, the relief of severe depression. This is why Dr. Insel and other neuroscientists are so excited about ketamine, which appears to potentially be a breakthrough treatment, working quite differently than the traditional SSRI antidepressants.

Such scientific evidence is essential, of course, and the evidence for ketamine was exciting. But entrepreneurial doctors quickly rushed far ahead of any evidence. Just a whiff of excitement, with the barest shreds of scientific support—just as in the heyday of Neurontin bipolar treatments—and hundreds of new ketamine clinics popped up to provide a drug for widespread off-label use. Besides prescribing it for treatment-resistant depression, the ketamine docs quickly began promoting it to rapidly reduce suicidality in emergency rooms and for *maintenance* treatment of depression, for which there is essentially no evidence. Not only that, but it was also prescribed for the treatment of obsessive compulsive disorder, for post-traumatic stress disorder, for chronic pain, for migraine headaches, for addiction, and for innumerable other conditions—all in the absence of data. The newly organized American Society of Ketamine Physicians soon listed over four hundred clinics across the country providing these off-label ketamine infusions, many of them run by anesthesiologists or other nonpsychiatrists. They were accused of not thoroughly screening patients, of offering the drug to anyone who can afford it (at a cost of $350 to nearly $1,000 per treatment, it is generally not covered by insurance), and of poorly coordinating care with the patient's own mental health providers.

And ketamine's risks are far higher than those of Neurontin. Ketamine can produce euphoria, dissociation, and hallucinations, and it has a significant risk of abuse. Since ketamine's patent expired long ago, it's doubtful that the FDA will *ever* give it an official "indication" for depression or these other disorders. A new formulation, nasal esketamine, brand name Spratavo, consists of the left-handed molecule of ketamine. Spratavo *was* recently approved by the FDA (our DES research group was briefly involved in one of these studies). Since Spratavo costs $5,000 to $10,000 per course of treatment, it is extremely unlikely to put the ketamine clinics out of business.

So now, more than ever, we are living in an off-label Wild West. Will these new off-label uses lead to benefits or catastrophe? Will they be major advances, or will they lead to unnecessary suffering and deaths?

Which leads to inescapable conclusions: our system for evaluating and regulating off-label uses is broken. And we need a better system all around.

Obviously, it's essential to curb the drug companies' abuses. After the Neurontin litigation, the companies did significantly mend their ways. They began separating the drug reps, who give drug samples to doctors, from the "MSLs"—medical science liaison staff—who give information about the drugs. There is now a legally mandated firewall between drug salesmen and drug information providers.

Also, hospitals began making rules to limit drug companies' influence—banning free lunches, free doctor bags for medical students, and so on—and the FDA began requiring annual reports of all payments to doctors through the Physician

Payments Sunshine Act, federal legislation passed in March 2010 as part of the Patient Protection and Affordable Care Act. Like most doctors, I now refuse freebies, meals, outings, and payment for speakers' bureaus. All for the better.

SEEKING TRUE SIGNALS

The thing is, just stamping out off-label promotions misses a bigger point: exploring off-label uses can be incredibly *valuable* to medical care. Even revolutionary. When and if the data finally come in, ketamine may indeed revolutionize the treatment of depression, just as azidothymidine (AZT), an orphaned cancer drug that didn't work for cancer, was reborn as a revolutionary treatment for AIDS. Because such amazing breakthroughs may lurk within the garbage dump of existing drugs, it makes sense to actively *seek out* off-label uses. Even case reports and small studies of off-label uses, which commonly appear in obscure medical journals, may signal possible benefits. Even more strikingly, off-label prescribing patterns frequently emerge in the absence of research. Most uses are likely foolish, even dangerous, but some of these uses *can* result in significant advances.

We need a better system to seek out and test such uses. This search, I would argue, is too important to leave to serendipity. In 2012, the National Institutes of Health began one such initiative, the National Center for Advancing Translational Sciences (NCATS)—to search for new uses for existing drugs. Inspired by the success of AZT for AIDS, NCATS tests drugs that have been abandoned by pharmaceutical companies *before* they entered the marketplace (because they failed to show effectiveness for their initial purpose) and provides funds to investigate other possible uses. Already, NCATS-sponsored studies have identified several possible new anticancer drugs, along with a possible anti-Alzheimer's treatment.

NCATS is a good start, but it is limited—after all, it only applies to drugs that *failed* to get FDA approval. What about drugs that are already out there in the marketplace? Many drugs, in a sense, are mislabeled from the start, since their initial use has been chosen based on commercial expediency. When other possible benefits emerge later, there's often no money to explore them. They've been unnecessarily sidelined, even abandoned.

One way is to require additional studies when an already-FDA-approved drug starts to be used off label at a high rate. These studies need to be done properly, especially since off-label uses are associated with more side effects and toxicity. If the drug is still under a patent, the manufacturer should clearly pay for them. If the patent has expired, as in the case of ketamine and a host of other drugs, they could be funded by a combination of drug company moneys and federal dollars, pooled so that no one company would have undue influence on the findings. Perhaps a portion of profits from proven new uses could go back to support future studies.

These steps would bring us a long way toward optimizing the benefits of off-label uses.

But the biggest things we need, in my opinion, are humility and an openness to wonder, neither of which usually come to mind when you think of the pills in your medicine cabinet.

Just think of it: when a drug—any drug—is approved by the FDA and introduced to the marketplace, the truth is that we know amazingly little about it. In a way, that's no surprise. To get FDA approval, studies need to prove that the new drug is relatively *safe* and to demonstrate that it works for one particular illness, but that's about all. We may still be blindsided by serious toxic effects (Thalidomide, for instance) that don't emerge until a drug is in the open market. Often, we don't even understand *how a* new drug works (Neurontin being only one of many examples).

Even more frequently and, in a way, wonderfully, we have no clue of a new drug's *best* use. This is especially true for psychiatry in the emerging age of neuroscience, and this is the area of most interest to me as a clinician-scientist. This is where the sense of wonder comes in. And here is my bias: I feel guilty love for off-label uses. I need them every day for my practice as a psychopharmacologist because I see many people with "treatment-resistant" disorders, for whom approved drugs haven't worked sufficiently, yet at the same time I feel uneasy for choosing drugs with inadequate proof of effectiveness. Beyond that, as a researcher, a clinical trialist, I find them intriguing for their illumination of scientific questions.

Viagra, for instance, was initially developed as a cardiac medicine, for the treatment of high blood pressure and angina pectoris—severe pain in the chest caused by inadequate blood flow to the heart—but when it was being studied for angina, the patients reportedly wouldn't return leftover pills to their doctors (a general requirement of FDA studies). It turned out that the study participants were hoarding the pills because they had other, fairly amazing, effects. While intended to dilate blood vessels in the heart, instead Viagra dilated blood vessels in sexual organs, causing welcome erections and improved sex for the mostly middle-aged male research subjects. Essentially, their "diversion" of study drug samples led to an understanding of the importance of the phosphodiesterase enzyme (PDE) in sexual arousal, which in turn led to the development of an entire new *class* of drugs—PDE inhibitors—including Viagra, Cialis, and Stendra, for what is now called "erectile dysfunction." Oh, and there's another twist: it turns out that Viagra also helps dilate blood vessels in the lungs. Viagra now has an FDA indication for the treatment of pulmonary hypertension. Sort of an *off*-off-label use, two generations later.

The greatest frontier in off-label use is in the brain. Take the blood pressure medicine Prazosin. What is the principal effect of Prazosin? By blocking alpha-1 receptors in muscle cells surrounding our blood vessels, it widens the vessels, decreases resistance, and improves blood flow. But Prazosin also can get into the brain, crossing the "blood-brain barrier." Alpha-1 receptors appear to play a significant role in the brain in severe stress, particularly in people with post-traumatic stress disorder. Prazosin can be quite effective in suppressing nightmares in PTSD. It has changed the lives of many patients traumatized by the events of September 11, 2001, and leads to intriguing questions about the neurochemistry of trauma.

Then there is the existing drug methadone, which has been used in the United States for the treatment of addiction since the 1960s. Given once a day, it has enabled millions of heroin addicts to stabilize their lives and work toward recovery. A worthy drug. Yet recent research has shown something at once amazing and unsurprising: methadone is actually *two entirely different drugs*! Like many drugs in current use, methadone is a "racemic" mix of left- and right-handed molecules (think of scissors, which can be customized to the handedness of their users). Based on handedness, molecules can have very different properties. Sugar, or glucose, also called dextrose, is a right-handed molecule (*dexter* is Latin for right). Its left-handed mirror image does not appear in nature. The miraculous Parkinson's disease drug L-dopa is the left-handed molecule; the right-handed dopa molecule, D-dopa, is poisonous. Methadone's left-handed molecule is an opiate and is what accounts for its success in heroin addicts. In contrast, the right-handed methadone molecule is an NMDA-receptor antagonist. It is apparently not addictive and does not have opiate properties. It is currently being studied as an antidepressant medicine and may have rapid onset of action. Since it has a different "mechanism of action" than existing SSRI antidepressants, it may be a welcome advance that was sitting right in front of our eyes for decades until smart chemists did their homework!

Even more intriguingly, new studies show that the diabetes drug Metformin may have strange, almost unprecedented effects. It may actually slow the cellular aging process. Unsurprisingly, in the absence of high-quality data, it has already become widely used off label by life-extension enthusiasts. But *does* it slow the biological process of aging? I'm skeptical about this one, but truth to tell, I can't wait to see what the studies show.

Hence my excitement after all these years of practicing as a doctor whenever a drug with a new "mechanism of action" is introduced to treat some illness, whether psychiatric or medical. Or when an old drug, long taken for granted, is looked at anew with advanced scientific tools. Sure, this drug, old or new, may help with a particular disease, but what *else* might it do? Whom else in my practice might it help? As a researcher, I'm eager to discern a true if often faint signal amid much noise. Or to help prove that the drug, despite massive hype, actually *doesn't* work.

The risks of off-label prescribing are now painfully obvious, thanks to Dr. Angell and her fellow muckrakers. But there are also virtues of off-label use, as a means of exploring our world and curing disease. Off-label uses can be miraculous or catastrophic. At worst, they are dangerous, sometimes despicable, even criminal, and potentially lethal. At best, they are a means by which we—doctors and patients—can explore the human body, the human psyche, and the human brain. Rigorous studies, high-quality investigations of off-label uses, can advance the treatment of disease and illuminate the unending mysteries of human biology, guiding us into the unknown. And in the Age of the Scanner, off-label uses, by helping us explore the mind and brain, can potentially lead to effective new treatments of disabling disorders.

CHAPTER 13

Mind Wandering, Then and Now

New Views Over Three Eras, 2005–2023

CAPTURING DAYDREAMS

As a kid in the 1960s, I spend endless hours staring out of windows. Bored in class, my desk piled with mimeographed worksheets that are invariably sent home never to be completed, I lounge in the back of Mrs. Wilson's third-grade classroom of Roxboro Elementary, daydreaming away. And at home in my attic room, I page through Superman and Green Lantern comic books and find my thoughts wandering out the window again, heading wherever they might, above the streets and houses and trees of our quiet Midwestern neighborhood, my X-ray vision seeking criminals and cosmic villains.

Years later, as a fledgling neuroscientist at Columbia—still prone to daydreaming in a dull lecture (my wife diagnoses me with inattentive type of attention deficit disorder)—I am hardly surprised to learn that daydreaming is an incredibly active brain state. Just close your eyes, let your thoughts meander, and the medial temporal lobe starts to chatter up a storm with the prefrontal cortex and the posterior cingulate cortex.

But why? What's the purpose of such mind wandering? Perhaps, say scientists, to generate ideas, integrate thoughts with experiences, and consolidate memories. Whatever its purpose, it seems essential to healthy brain functioning.

Reading about MRI scanning, I am intrigued to learn that you can image the brain's daydreams, or at least the circuits involved. A few minutes in the MRI—with the instruction "Relax and lie still"—can activate the brain's daydreaming circuitry, what has been called the Default Mode Network, or the DMN. It's a robust network, found in every species of animal, and it's hundreds of millions of years old. It is during rest that the DMN activates—when a cat imagines catching mice or a bored third-grader becomes Spiderman. In all species, the DMN quickly deactivates when there is a need to act in the world.

Not only is the DMN essential to mental functioning, but its activity is abnormal in many illnesses. In schizophrenia and autism, DMN hyperactivity may lead to such an excessive internal focus that people find it impossible to switch focus to the outside world.

The DMN may also go awry in mood disorders. Depressed people's minds often flood with fear, worry, self-criticism, and suicidal urges. MRI scans of depressed people show increased crosstalk between the ventromedial prefrontal cortex, the anterior cingulate, and the lateral temporal cortex—brain areas related to "self-referential processing" and negative self-ruminations. These circuits are overactive and hyperconnected in depression. Yvette Sheline, from Washington University in St. Louis, found that depressed people have difficulty deactivating their DMNs when trying to solve problems or to pay attention to others. In a sense, depression may be an inability to *stop* mind-wandering!

But I'm getting ahead of myself—first I need to tell you how I was lucky enough to capture daydreams. It started several years after I came to the New York State Psychiatric Institute, when our new MRI scanner was installed, and a call came out for new studies. In chapter 10, I described how Brad Peterson and I were able to set up a study combining a double-blind clinical trial with repeat MRI imaging. Since we were going to the trouble of starting a complex study, Brad advised that we get as many types of scanning sequences as possible, whatever we could obtain over the patient's two-to-three-hour session in the scanner. We would gather our data, and later we could figure what kinds of analyses to run.

We first study the medicine duloxetine, brand name Cymbalta, a newly introduced serotonin-norepinephrine reuptake inhibitor, or SNRI. Once that project ends, we are fortunate enough to get funding for an almost identical study of another new SNRI, desvenlafaxine, brand name Pristiq. Eventually we end up with MRIs on over one hundred subjects, from before and after they receive either antidepressant medicine or placebo.

Our images include anatomical scans that can define brain structures as small as a few millimeters in size. We also have "diffusion tensor imaging" (DTI) scans, which measure the connectivity of one part of the brain with another. We have "magnetic resonance spectroscopy imaging," or MRSI scans, which will allow us to quantify various brain chemicals, or neurotransmitters, throughout the brain. We obtain pictures of cognitive circuits by asking our patient to solve puzzles, and we use vivid photographs to capture emotions in the brain.

Finally, by using "resting state" imaging sequences, so-called rs-MRI, we measure the activity of half a dozen brain networks. Including, it turns out, the DMN.

With such a treasure-trove of images to analyze: where to start? We enlist a corps of grad students, postdocs, and biostatisticians to data clean, process, and analyze. On anatomical scans, our images show how antidepressant treatment alters brain *structure*. On MRSI scans, they reveal how medication normalizes levels of brain *chemicals* including glutamate, glutamine, and N-acetyl-aspartate. Using machine-learning techniques, they show how brain networks, nodes, and hubs are remodeled during the return to normal mood.

Amazingly enough, an offhand hallway chat with Brad Peterson leads to cutting-edge science. Our findings are eventually published in top journals including *JAMA Psychiatry*, *Lancet Psychiatry*, and *Molecular Psychiatry*. It is all rather mind-boggling, especially since, to put it mildly, I have never seen myself as much of a scientist. (During my years of rebellion in my twenties, I always swore that I'd never do research!)

Because our studies compare medicine to placebo, we can be reasonably sure that all such changes come from the treatment itself, not just the effects of time. Some findings surprise us: we have predicted that the brain's cortex will get thicker with antidepressant treatment. Instead, our chronically depressed patients' cortices get *thinner* after treatment, resembling the brains of people without depression. Perhaps the long duration of depression in our patients—a decade or more—has thickened their cortical layers to compensate for the overactivity of lower brain regions like the amygdala; in contrast, briefer depressions may cause thinning.

We are surprised in one study to find that brain systems related to *pain* show significant change after antidepressant treatment. "Depression hurts," our patients often say, and suicide notes commonly contain some version of the phrase "I can't take the pain anymore!" Our MRIs reveal that this may literally be true. Depressive agony may reflect overactivity in pain-processing brain centers.

No doubt our MRIs will reveal other secrets. For instance, I'm dying to look at the hippocampus—the elegant, whorled, seahorse-shaped deep center of new memories, of mapping the world—to see how its structure and functioning change with medication treatment. If only we can get funding to do the laborious pixel-by-pixel analyses of several hundred images, Brad and I could investigate the profound question of whether the depression-injured hippocampus can regenerate!

But I am most proud of our findings about mind wandering, our 2013 paper on the Default Mode Network, the DMN, published in *JAMA Psychiatry*. Dr. Jonathan Posner, a brilliant young child psychiatrist and brain imager at NYSPI analyzes our resting-state MRI scans, rs-fMRIs. His analyses confirm that DMN activity *is* increased in our patients with chronic low-grade depression—just like what is found in more severe major depression. Not only are their DMN circuits hyperactive compared to healthy people, they are also "hyperconnected," suggesting these brain patterns have been present for many years.

Intriguingly, after antidepressant treatment our patients' DMN activity level drops to normal—the same level seen in people *without* depression. Whereas, for those given placebo in the study, DMN activity does not budge.

Have we found an instance of what psychiatry in our neuroscience-obsessed era is seeking with almost religious zeal: an illness "biomarker"—a reliable measure of brain activity gone awry—that can normalize with treatment?

It's easy to get caught up in the media attention our paper gets after publication, but we try to be cautious. These findings require replication; the methods are new and not entirely proven. But still, I am psyched: look what the scanner can do in our new age of neuroscience! Harking back to my Action Comics days, I think of it this way: *we have ridden the MRI scanner to new frontiers, traveling deep through the mysterious human brain!*

And look what can be accomplished with the right study, the right tools, the right team.

MAGNETIC TRANSFERENCE IMAGING

And of course, in thinking about mind-wandering deep in the bore of the MRI, my thoughts return to training days—the early 1980s—when mind-wandering is essential to an entirely different treatment approach.

Back then, before the internet, the Kardashians, Bitcoin, and YouTube, all supervisors worth their salt—even on inpatient units—needed to be psychoanalysts. Back then, every therapy office had a couch and generally two ashtrays, since the wafting of smoke seems essential to freeing the mind to roam. "Free association," we call it. Free association is central to the mindset of psychoanalysis.

Where does your patient's mind go when you sit with her? my long-term psychotherapy supervisor Dr. Banks asks me in those now-demolished offices of the Payne Whitney Clinic, in a now wholly mythical world. We are using that century's once cutting-edge (but now well-worn) technology to explore mental processes and find a way back to sanity. And, at times more uncomfortably, where does *my* mind wander?

Free association is a "fundamental technical rule of analysis," Freud proclaims in his 1917 *Introductory Lectures on Psycho-Analysis*:

> We instruct the patient to put himself into a state of quiet, unreflecting self-observation, and to report to us whatever internal observations he is able to make. At the same time, we warn him expressly against giving way to any motive which would lead him to make a selection among these associations or to exclude any of them, whether on the ground that it is too disagreeable or too indiscreet to say, or that it is too unimportant or irrelevant, or that it is nonsensical and need not be said.

To my 2020s ear, our current-day instructions to patients in the MRI scanner—*close your eyes, let your mind wander*—eerily echo Freud's instructions of a century ago. Yet we now proceed with a different aim: whereas once we asked the patients to calm their minds so their insights could bubble up from the disinhibited mind, now we instruct them to lie still so we can probe their neural circuits and gather insights for ourselves.

Back then, in my Payne Whitney days, the patient lounged on the psychoanalytic couch in a state of awakened dreaming, the analyst listening with his third ear, only occasionally speaking. In the *DSM* era—an age of standardized diagnoses and data-based therapy and medication treatments—we listened just as acutely but checked off symptoms and assigned diagnostic labels, all to prescribe new medications and psychotherapies. Now—in the era of the new neuropsychiatry—laboratory insights guide our treatments. Our new transference object is the MRI scanner itself, and our eager patients are swaddled in cotton blankets, heads immobilized, as a motor slides them smoothly inside a sleek metal tube. Goggled up, mouse pad or toggle box taped against their thighs, they merge with the machine and gaze intently at images or click the mouse—or let their thoughts wander.

And, in the new neuropsychiatry era, we capture the mind at work—synchronized activation patterns of various brain networks. We trace aberrant circuits, abnormal activity, or sluggishness in different brain regions. We let our own minds soar—envisioning ways to make these circuits healthy again.

Now I supervise young doctors in training, third- and fourth-year psychiatry residents, researcher-clinicians—we are all charting our way into the new era. These young doctors probe and scan; they swab cheeks for DNA samples; they insert fluorescent genes into lab animals to light up appropriate brain circuits in bright orange and green; they inject radiolucent markers to track neurotransmitter levels in the PET scanner. They have an uncanny knack of envisioning the psychiatry of the future. In myriad laboratories at the institute, they develop animal models of "discounted decision making"; they clone genetically fearful mice; they grow "schizophrenia in a Petri dish," things that seem unimaginably strange, which they then relate to the psychic suffering of the patients in the clinic.

I have amazing conversations with these neuroscience-savvy trainees. We swap insights into brain aging and how it might be slowed with brain stimulation; we talk about telomeres, the shoelace ends of our DNA, which are shortened by stress, and consequently, how our patients should learn to better manage stress, using mindfulness meditation and yoga. We talk of modulating the activity of the DMN—with n-acetylcysteine, an anti-inflammatory amino acid derivative.

We talk of enhancing, through regimens of physical exercise, neuronal growth factors such as brain-derived neurotrophic factor (BDNF) that improve brain health. Then we start regularly prescribing exercise to our patients (and ourselves!). We speak of cytokines and other chemicals related to inflammation that are associated with brain atrophy: how can we help our patients decrease systemic inflammation and thus improve outcomes of depression and other disorders?

We discuss the intriguing evidence for *regional aging* of specific brain centers in psychiatric disorders—how anxiety disorders can result from parts of your brain aging at an accelerated rate. And, more optimistic, we discuss the possibility of enhancing "positive neuroplasticity" as our patients (and we!) grow older.

These are entirely new conversations, so different from the ones I had with Payne Whitney Clinic supervisors back in the 1980s, or with young doctors in the 1990s in my days at Manhattan Medical Center, or even from the ones we had here at NYSPI just a decade ago.

Unsettling, sometimes mind-boggling evidence that a new era has arrived.

THREE WAYS OF LOOKING AT A PATIENT

Lena is twenty-six, well-schooled, a classic failure to launch. More accurately, having quit her job as a corporate lawyer after two miserable years and bunking at her parents' home, she's more of a boomerang case. Beset by depression and severe panic, she wastes days enviously Insta-stalking her college roommates, who have jobs, co-ops, chunky engagement rings. Their lives move ahead, whereas a decade of depression has sent her tumbling backward. Her sleep sucks, she tells me, she stays up endless hours, then crashes much of the day. Her eating habits, never good, have worsened dramatically; she has had a return of her old habits of self-starvation and food restriction, which had become full-blown anorexia during college and now threaten again. She ghosts friends until they give up on her; she rarely ventures out. She ruminates endlessly, her mind wandering recursively over the same burnt-out territories of misery and self-blame. People praise her strengths: her keen wit, her intuitiveness and creativity. Praise merely sharpens her pain and enhances self-flagellation: knowing the hopelessness of her future, she Googles methods for painless suicide, which alone will bring surcease.

Back in my Payne Whitney days, in the 1980s psychoanalytic era, the treatment approach is clear: "Tell us your dreams," every therapist coaxes. "What comes to your mind?" Psychoanalysis, or at least psychoanalytic psychotherapy, is our tool. We have the couch, the fifty-minute hour, and the power of associative thinking. The 1970s and 1980s version of Lena searches for an analyst with availability and books three to four hours per week for the foreseeable future. Once arrangements are made, Lena will sit or, better, lie on a couch and let her mind wander, and in a state of awakened dreaming she will discourse on whatever may emerge.

Dream analysis, free association, and the development of transference toward the analyst: they are the tools. Her healer listens silently, save for the occasional pithy interpretation, showing infinite patience. They meet several times a week for as long as it takes: two, three, four years or more. The black ruminations may persist, though, even worsen. Indeed, an improvement of Lena's symptoms is far *less* important than "working through" her basic issues of attachment, aggression, and intimacy and her relation to primary "objects," especially her ambivalence toward her mother, who, no doubt, has failed in many ways. Her food restriction, bordering on anorexia, is clearly a measure of her fear of sexuality, her desire to regress to childhood, indicating how much work needs to be done. In fact, many psychoanalysts believe that recovering too quickly from depression might sap

Lena's will to delve deeply, to continue "the work." It may even be better that she *stay* depressed until the work is done.

Later, in the *DSM-III*, *IV*, and 5 eras, from the mid-1980s until the early 2000s, we perceive Lena's mind wandering far differently. This woman, this gifted young attorney, clearly suffers from major depressive disorder (MDD). Her sleeplessness, her low mood, her anxiety, her loss of appetite, and her persistent ruminations with self-critical and suicidal preoccupations are classic "A. Criteria" for that illness. To receive the MDD diagnosis, Lena merely needs to meet five or more of nine criteria. Her obsessive negative thoughts may be categorized under item 7, "feelings of worthlessness or excessive or inappropriate guilt . . . nearly every day," and possibly item 8: "diminished ability to think or concentrate, or indecisiveness, nearly every day." She likely also "meets criteria" for a diagnosis of anorexia nervosa, which sometimes cohabitates with depression.

The best treatments for her depression? Antidepressant medication, preferably one of the serotonin reuptake inhibitors, Prozac or Zoloft, taken on a daily basis for at least six to twelve months. Every month Lena will see her psychopharmacologist for "med checks." Psychotherapy can help too, especially "evidence-based treatment" such as cognitive behavioral therapy (CBT) or interpersonal psychotherapy (IPT). CBT's goals are to address Lena's negative thinking and avoidant behaviors, to encourage her to "challenge" or interrupt her "automatic negative thoughts" and to increase her "exposure behaviors"—pushing herself to go out and see friends, to start applying for jobs, to consider what kind of legal work suits her best. CBT can also help her set goals for healthy eating. Once-a-week CBT sessions for several months are recommended to start, along with weekly group therapy for her eating disorder. Once Lena's symptoms improve, the individual sessions may be cut back to every two to four weeks. Her prognosis is excellent, so long as she continues taking her SSRI.

Now in the Age of the Scanner, what do we see?

As I sit across from a patient in my office—this young corporate attorney, this college student, this older man—I wonder: *which circuits are amiss? Which need retuning, strengthening, muting?* In our era of the New Neuropsychiatry, so many disorders are now thought to result from overactive and hyperconnected brain circuits. For Lena, "early adversity" of her parents' dissolving marriage and her father's severe alcoholism led to enhanced "fear conditioning" and to activation of circuits that were adaptive at the time, making it easier to survive stressful events that bordered on abuse or neglect, but that now are maladaptive. Over time, she had "reconsolidation" of these negative memories and thoughts, which were etched into her brain by "epigenetic" changes, either via methylation or acetylation of segments of her very DNA.

Continual self-criticism initially enabled her to study harder, perform better, to excel . . . up to a point. Also, eating was one thing that she could control, and food restriction initially made her feel better. Now, though, her brain is locked into patterns that are extremely rigid and sometimes impossible to interrupt. Behavioral patterns have escaped her control and now rule her life. Lena's dark fantasies? Her hellish dreams? Circuits run amok, in need of retuning. Her DMN initially became hyperactive after her father abandoned the family and her mother was diagnosed with uterine cancer, and her coping style of trying to be a perfect student, to squelch feelings, to need no one, to exercise and diet obsessively, while initially helpful, eventually led to aberrant activity of her fear circuits. Her eating behavior led to out of control "habit circuits," which may endanger her life.

Sure, when seeing Lena, I still have *DSM* diagnoses in mind. I still write "major depression, recurrent, moderate severity" or "anorexia nervosa, restrictive type, without bulimia" on an insurance form when necessary, but increasingly the treatments I propose go *across* the particular *DSM*-5 categories. And it's fair to say that neither do I entirely ignore what I learned at Payne Whitney about psychodynamics and the unconscious mind.

But I see Lena with new eyes: as one of many New Neuropsychiatrists—a devotee of the emerging practice of neuroscience-based psychiatric practice. There are thousands of us neurobiology watchers, who follow the latest advances from scientific research at grand rounds and professional conferences and in top neuroscientific journals, a whole army of psychiatric pioneers who try to apply them wisely in our work with patients. We learn what affects the brain's activity, whether in sea slugs or mice or chimpanzees—then dare to make the leap to human suffering.

Some neuroscientists, including the former National Institute of Mental Health director Dr. Thomas Insel, insist that as yet neuropsychiatry has brought no advances to psychiatric treatment. In terms of specific repair of abnormal genes, sure, I agree, we are far from definitive care. I can't put Lena in a scanner and pinpoint exactly which circuits in her brain need retuning; I only have a general model of her brain activity. Instead, the New Neuropsychiatry brings a broad new perspective, a vision that incorporates cutting-edge advances in understanding brain activity—and interactions with physical health—into the treatment of psychiatric disorders with real patients. Instead of specific cures, we have useful principles.

But does the New Neuropsychiatry represent an *advance* over previous models? As a practicing psychiatrist and clinical researcher, when I compare it to the old psychoanalytic model and the more familiar *DSM* model, I am convinced that the neuroscience model *is* the most compelling of all and that psychiatry's progression from one model to the next is inescapable. The neuroscience model, the Age of the Scanner, brings psychiatry into the age of high science and into the mainstream of medicine.

Yet it comes with its own perils. There is a narrow neurobiological approach in which the brain is seen mostly as a complex squishy computer, a hive of virtual realities summoned by winding circuits, in which the only things of interest are those circuits and the mysteries of connectivity. In this view, the mind is irrelevant, trivial, an epiphenomenon. That narrow view is far too simplistic, too reductive. In my eyes, the new neuropsychiatry should be a broad and inclusive approach that can incorporate the mind, the brain, *and* the body and fold in the complex interactions between people in society that restructure our brains on an ongoing basis. This broader neuroscience seems to me to be an inescapable paradigm—and one that can't help but prevail over time. Overall, science must win out. At the same time, it is easy to get ahead of ourselves, to promise too much and then be disappointed. Many studies and findings of brain circuits and genes and treatments will be touted and then disproven by later research. Or more often, they will be found to be partly correct but better explained by new findings, by studies showing that things are more complicated than we initially thought. Complexity is disappointing but ultimately good and true.

The strengths of the new neuropsychiatry are undeniable. Innumerable studies have shown the importance of fundamental principles. Neuroplasticity is one such principle. Ongoing remodeling of brain structure and connectivity clearly occurs throughout life. It is also clear that psychiatric disorders can injure the brain, especially when untreated. Furthermore, it is clear that *medical* health has a profound effect on the health of the brain: diabetes, high blood pressure, obesity, and other medical conditions (not to mention habits like smoking and heavy drinking) can injure the brain and often worsen psychiatric outcomes. It is also becoming abundantly clear that the treatment of psychiatric illnesses, including medication, exercise, mindfulness, and other treatment approaches, has the potential to slow or even reverse brain injury and can enhance brain health. There is the key concept of "resilience," or positive adaption to chronic stress, a neuroscience-based concept, which can be a powerful strategy for the long-term management of disorders.

Such broad principles can be applied across many disorders and have positive public health effects, even if we don't yet have specific technologies for individualized neuroscience-based cures, even if we haven't reached our goal of "precision psychiatry."

Nevertheless, the New Neuropsychiatry has glaring weaknesses. Truly *individualized* approaches are frustratingly far off. Patients commonly come to my office asking if they can have a brain scan to find out what is wrong with them, the way that a cardiac catheterization could show what is wrong with the arteries of their heart. No, we're not there yet, no matter what Dr. Daniel Amen may write. We cannot yet do meaningful individual brain scans for most people. Also, to say something obvious on one level but not so obvious on another: mice aren't human. It's dangerous to generalize too enthusiastically from any one animal study to humans. There are immense distances *within* neuroscience, vast empty spaces separating findings from cell cultures in laboratories to lab rats and mice to humans. There are also huge risks of overpromising scientific advances based on small or flawed studies. We should expect that many initial findings won't be replicated when studied again with larger samples. Many negative findings should be expected if studies are done well, but there is a risk of becoming demoralized, of losing faith in a model that is basically sound.

The challenges of applying these new insights become clear when psychiatrists try to customize brain stimulation in the clinic. The brain circuits of most interest in psychiatric disorders are hard to get to with current technology, since they're deep in the brain. We are only beginning to do things like individualized focal brain stimulation for people with aberrant circuits: these treatments may be close but are not yet ready for prime time.

We also have a diagnosis problem with the New Neuropsychiatry. We have not yet carved out diagnosable circuit-based disorders that are applicable on an individual basis, so we psychiatrists are clearly not ready to replace the *DSM* in daily clinical work. Even if we did have these diagnoses, our existing medications are "dirty" and don't clearly target single circuits or neurotransmitters that may be identified.

Therefore, it's extremely difficult to provide *individualized* neuroscience-based practical guidance in the clinic, consulting room, or hospital. In this regard, the *DSM* is still much more user-friendly and useful. That said, neuroscience-based *principles* can usefully guide treatment, often in combination with approaches from previous ages.

So what would I recommend for Lena? We New Neuropsychiatrists do not entirely abandon the insights from previous eras, the Age of the Couch or especially the Clinic. We recommend effective medication and psychotherapy treatments—and SSRIs and cognitive behavioral therapy are certainly likely to help. Abundant research has shown that depression injures the brain and that it is essential to get it into remission to prevent further brain injury—and such evidence-based treatments may be essential to getting Lena's low mood, insomnia, and suicidality under control.

But beyond that, we increasingly are targeting circuits gone awry. Almost certainly we now recommend that Lena start mindfulness meditation or yoga to help her severe ruminations, her out-of-control Default Mode Network. Yoga can have a profound effect on the DMN and other brain circuitry. Experienced meditators have been shown to be able to decrease their brains' DMN activity at will, to turn down the activity of the fear-generating amygdala, and to switch smoothly from anxiety to relaxation. We might also recommend that Lena take a supplement called n-acetylcysteine (NAC), a compound with anti-inflammatory effects, which seems to decrease DMN hyperactivity. In a wide range of psychiatric disorders, including bipolar disorder, anorexia, OCD, addiction, and depression, NAC has been shown to decrease repetitive negative thoughts, to ease ruminations—whether about food or drugs or one's self-worth. (NAC is FDA approved for treatment of acetaminophen, Tylenol, overdose but is commonly used off label by psychiatrists; will NAC ever receive FDA approval for psychiatric illnesses? Hard to imagine, since there's nothing to patent, to capitalize on.)

We also will urge Lena to start a program of physical exercise, which should have profoundly positive effects on her mood and anxiety levels and on her brain circuits. Neuroscience research shows exercise to be an incredibly powerful way to decrease inflammation throughout the body and brain—it lowers C-reactive protein (CRP) and other inflammatory compounds such as the cytokines. It also boosts body chemicals related to the health of brain cells, including brain-derived neurotrophic factor (BDNF). We now realize that exercise is key to helping reverse brain injury caused by depression—shrinkage of neurons, decreased synapses, decreased "arborization" or branching, and increased cell death.

And we are certainly thinking about the long-term health of Lena's brain. Increasingly, we New Neuropsychiatrists are obsessed with this. It seems obvious that psychiatrists *should* fret about brain health, the way a cardiologist worries about the health of her patients' hearts. But oddly enough, and somewhat mysteriously, neither the psychoanalysts (in their often-smoky offices) nor the *DSM* psychiatrists (in their busy clinics) paid much attention to the health of the very organ that causes psychic distress.

Whereas in the Age of the Scanner, brain health is of primary concern (not only our patients' brains, I might add, but our *own* brains and those of our loved ones). Not only can depression and anxiety injure the brain, not only do our treatments have the goal of preventing further injury, but increasingly these days, our treatments also have the goal of inducing what we call "positive neuroplasticity"—a process of brain recovery and repair.

To that end, our goal in working with Lena—beyond getting her depression and eating disorder under control and getting her life back on track—is to help her increase resilience, her ability to cope with life stresses over time. Resilience is defined as the process of positive adaptation in the face of adversity, trauma, tragedy, and other threats to well-being. People can learn to become more resilient, to deal better with adversity, and even to thrive under stress. From a scientific point of view, resilience is complicated, the result of a host of biological, social, cultural, and psychological factors. For a New Neuropsychiatrist, these are all researchable areas worthy of active study, which makes me optimistic that we will develop ever more effective and more specifically targeted treatments.

And if all these gradual approaches fail to help Lena, if her depression and anorexia turn out to be severely treatment resistant? Then we will increasingly start thinking about *radical* resets of neural circuits, treatments intended to disrupt the abnormal brain activity of "dis-order" and setting the stage for a process of healthy reconnection. Break the habit circuits, disrupt the broken-record DMN, reset the cruel circuitry of addiction: we have our psychotherapy approaches that are helpful but often laborious. Cognitive behavioral therapy for depression, exposure therapy for OCD, dialectical behavioral therapy in borderline personality disorder. They help but take much time and effort. Medications too, SSRIs particularly, disrupt these closed loops, but they too, like psychotherapy, often require ongoing treatment, years, even decades.

Maybe—we New Neuropsychiatrists increasingly wonder these days—there are ways of speeding response. Ways to introduce healthy *entropy* into these closed-loop systems of disorder—a reset, a reboot of sorts, something that would be beneficial, safe, effective, quick. But how?

Transcranial magnetic stimulation is one promising option. A zap of magnetic stimulation can potentially be directly targeted to aberrant brain circuits—if we can find the correct target and can burrow far enough through the skull to affect these often deeply buried exchanges.

Intravenous ketamine, which has proven to be a powerful antidepressant, at least in the short term, is another. Glutamate-like drugs stimulate the NMDA receptor, apparently reinducing plasticity into the rigidly disordered brain. Even old-fashioned electroconvulsive therapy, ECT, which has been modified to have less injurious effects on memory, may perform a rapid reset of such disturbed circuits.

A new combination of two old drugs also may lead to rapid response: the old cough medicine dextromethorphan, when combined with bupropion, appears to quickly improve depression by boosting activity at the NMDA receptor—one of the strangest off-label uses I've ever heard of! Who could imagine that an over-the-counter cough medicine, introduced in 1958 and a component of some formulations of Robitussin and Mucinex, would be "an NMDA receptor antagonist that acts as a SERT and NET inhibitor, nACh α4β4 antagonist, sigma-1 and mu opioid receptor agonist"—and discovered to have potent antidepressant effects more than half a century later?

Intriguingly, we are taking a new look at the long-maligned psychedelic drugs—LSD and psilocybin and MDMA, which bind powerfully and weirdly to the $5HT2_A$ serotonin receptor system, a system widely distributed throughout the prefrontal cortex. These compounds set off a host of strange hallucinatory and sensory and cognitive phenomena and are reputed often to lead to psychic healing. Do psychedelic drugs too work by rapidly resetting the dysfunctional circuits of psychiatric illnesses, by (as some proponents claim) inducing a process of brain *hyper*plasticity? And if so, is such hyperplasticity a good and healing thing?

So that is why, these days, when I sit in my Upper West Side therapy office or when I see research patients at the New York State Psychiatric Institute, I am subject to strange visual experiences. Like my thousands of New Neuropsychiatry colleagues, I can't help but see *through* my patients' skulls. My years of MRI gazing, hundreds of views of mind-wandering circuits, not to mention my childhood obsessions with DC and Marvel Comics, have given me X-ray vision.

A graduate student talks about how he can't sleep after being mugged on the subway last month; a middle-aged graphic artist can't relax since her wife walked out; a software engineer is tortured by solutions for his failing startup's data systems problems that keep rattling through his mind . . . As I gaze through their skulls, I can't help but visualize my patients' brains at work. Aberrant default mode networks, leading to out-of-control negative ruminations. Difficult-to-activate cognitive-control networks, making it impossible to set priorities at work. Hypersensitive amygdalae, making it impossible to experience a sense of safety. And sluggish dopamine-based reward circuitry, leading to low motivation and an inability to get started on important projects. I see—more precisely, I imagine—brain processes that make my patients unable to respond effectively to opportunities and rewards of daily life.

Half a millennium ago, a similar phenomenon occurred. In the days of the great sixteenth-century Dutch anatomist Andreas Vesalius—who first dissected our human anatomy, organs and bones and nerves—practicing doctors were startled by their novel ability to visualize the flow of blood through their patients' hearts and great vessels, the aorta and vena cava. They were amazed to perceive the breath of life flowing from a patient's mouth to pharynx and down into their ever-branching bronchi and alveoli.

Now, we neurobiology watchers, we searchers and scanners, are equally amazed by new powers. New Neuropsychiatrists all, we begin to peer through our patients' skulls, bearing our invisible MRI spectacles, our Google Glass NeuroExplorer Edition goggles. On the subway, in the fuselage of a 737, or in a hushed consulting room, we are entranced by our vision of electrochemical signals whirring through the brain circuits of the guy or lady sitting a few feet away.

And if we're being good doctors, good neuropsychiatrists, we also see our patients as true and unique individuals. It is not that we *still* see them this way, despite the progress of neuroscience. No, we see them as *more* miraculously individual than ever *because* of the progress of neuroscience. That's the wonder of our new era: we now know with certainty that each person has a unique brain with billions of circuits sculpted by unique life experiences, by individual losses and traumas, by their own chosen actions and evolved habits, by hopes and fears, and these thoughts and experiences and behaviors have fingerprinted unique epigenetic changes into their very DNA. And we know chemical brain patterns can cause suffering and disorder but can potentially be fine-tuned to bring people back to health, if only we can come to understand them well enough.

We humans are all of a kind, we realize. And, more profoundly than ever, we know that no two of us are alike.

THE IMPOSSIBILITY OF REASON

Here's the thing, though, literally keeping me up at night. What do *I* do with this new vision? As a practitioner, the experience is often inspiring, leading me to seek new ways to help my patients, enlarging treatment options, and undoubtedly improving outcomes. Exercise, mindfulness, meditation, retraining circuits, overwriting habit circuits by creating new habits. My patients seem to benefit, and scientific papers support what I see.

But as a researcher this new vision is daunting, inspiring, and overwhelming. As we enter the second decade of our new millennium, my research group, the Depression Evaluation Service, faces an existential crisis. It's ironic, since we psychiatrists now have such amazing tools and are tantalized by a host of exciting new research questions, but the world of clinical psychiatry research has been shaken up irretrievably. Our clinical trials expertise in the Depression Evaluation Service was central in the era of the *DSM*. We were founded in the late 1970s and have been part of the major psychiatric studies of depression for decades, for example, the 2001 STAR*D study, which compared different treatment options for treatment-resistant depression, and, more recently, the 2011 EMBARC study, in which my friend and colleague Pat McGrath MD was a key investigator, which looked for "signatures" of antidepressant responses using MRI, EEG, a host of neuropsychological tests, and which after five years is coming to an end.

But now? A decade and a half into the new millennium, we in DES are no longer at the white-hot center of psychiatric research. In the DES, we lack skills in optogenetics and PET imaging and cognitive science, all the hot new growth areas. In 2013, the National Institute of Mental Health announces—amazingly enough—that it will no longer provide *any* funding for traditional clinical trials! Every new study must be based on brain circuits, by a hypothesized mechanism of action in the brain. But hardly any circuit-based *treatment* studies are being funded.

The large clinical trials with hundreds or thousands of patients that the DES pioneered are not even eligible for funding. Such studies should be "left to industry," we are told. But Big Pharma has (shockingly enough) largely abandoned psychiatric research, since there's more money to be had in cardiology or cancer drugs. Pfizer, Eli Lilly, AstraZeneca, GlaxoSmithKline, and others are actually shutting down their psychiatric drug research pipelines. One after another, *Kaput*!

Plus, my colleagues in the DES are aging out, taking retirement. It has been a good run, since the 1970s, over four decades of cutting-edge psychiatric research. Clinical trialist friends from dozens of other departments around the country are also closing shop.

The *DSM* research era is over.

Once the DES concludes our participation in the EMBARC study, Pat McGrath announces that he's not applying for any more grants. Jonathan Stewart is doing studies of bright-light therapy and wake therapy for depression, which are fascinating and creative but which no one—federal agencies or foundations—will fund.

"It's just you in the DES now, David," more than one of my colleagues says. "What are you going to do next?"

I lose a lot of sleep trying to find the answer. I've made transitions before; maybe, just maybe, I can find a place in the new world. After all, back in the early 1980s I was trained as a psychoanalyst at Payne Whitney; then in the 1990s (despite the most minimal of preparations) I remade myself as a *DSM* psychiatrist and a clinical trials researcher at Manhattan Medical Center. Now, in the new millennium, can I similarly make another transition—perhaps to become a neurocircuit researcher?

One answer, which I start testing: collaboration. "What are you going to do next?" "Collaborate." Soon it becomes my stock answer. "Connect with other groups, we can run the clinical parts of the studies, they can do the fancy imaging and genetics." Why not? My completed large clinical trials combined with MRI imaging are surely a good start: our papers from two nearly identical studies have appeared in top journals. Pain networks, cortical thickness, the DMN. The good old DMN has turned out to be central to so many disorders and to designing new cutting-edge treatments.

Dr. Jonathan Posner and I put in grant applications to further investigate the DMN and other brain networks as well. More medication studies, to start, then we hope to propose studies of meditation and exercise, which also have been shown to impact the activity of the DMN.

I put in grants with a group doing transcranial magnetic stimulation, trying to find whether specific brain circuits can be identified in a person, then target those circuits with a magnetic beam. "Precision medicine" is the idea. I work with a group studying bipolar disorder using PET scans; we get one small study funded but struggle to get additional funding. Our DES group takes part in studies of some of the few new drugs still being investigated in psychiatry: a preparation of ketamine that is delivered by a nasal spray, for one, and a hormone derivative that is given intravenously to help women with postpartum depression. Together with the director of the Breast Cancer Clinic, we propose a study for depressed women with breast cancer and submit it to the National Cancer Institute. That one, alas, is still in purgatory, not yet funded.

Then comes a health crisis: I am diagnosed with prostate cancer. It's a family thing: my dad was diagnosed with that illness, and also his younger brother, Earl, and one of my cousins. Luckily, my diagnosis is made early, and I undergo a successful surgical procedure, then an unpleasant and gradual recovery, and am determined on a series of annual follow-ups to be cancer free.

Maybe it's time to hang it up. Regardless of the promise of the Age of the Scanner, maybe for me as an individual, it's time to step back. To retire from my job at the institute, focus on my practice and my writing. Move back into the city, take it easy. I can treat patients, be a theorist, an observer, a teacher, I can blog, post, Tweet, Insta-whatever, write more books.

Perplexingly, however logical that may be, I am possessed by an utterly irrational obstinacy, a desire to stay in the game, both for myself and my area of clinical trials.

There's got to be a plan to make it work. It's crazy to throw away clinical trials in this new era of neuroscience, the age of brain circuitry. Our clinical trials expertise is hard won, I tell everyone, and will be extremely difficult to restart once shut down.

I keep making that argument, and no doubt my colleagues get tired of hearing it.

When stocks go up, you are tempted to sell all your bonds. I declaim. *But is that wise? Shouldn't you keep the bonds, perhaps even buy more bonds to rebalance your portfolio?*

My arguments, alas, sound increasingly like those from my last years at Manhattan Medical Center. And anyhow, maybe it's not a stocks-and-bonds argument, maybe it's automobiles versus buggy whips, internal combustion versus horses, maybe I'm just plain wrong.

Maybe it's an identity thing: my out-of-nowhere later-life fantasy of being a researcher, which I seem to have bought into in a big way. Now, a decade or so into it, I still want a part of this exciting New Neuroscience world. Still want, despite innumerable obstacles. You could say I'm ruminating about it: it's a new preoccupation of my Default Mode Network, daydreaming of new research directions, new opportunities.

Who am I kidding?

Why those daydreams now?

CHAPTER 14

Floating Brains and Magic Mushrooms

Ancient Psychedelics Test the Progress of Psychiatry, 2019 to Today

ANOMALIES

"Thanks for meeting with me," Sander says, taking a seat in my new office.

I moved here just a few months ago, when Pat McGrath retired after nearly forty years at the New York State Psychiatric Institute. I've taken over as director of the Depression Evaluation Service—a daunting prospect, since it's nearly impossible to get research grants these days.

My new office, formerly Pat's, has a spectacular view of the Hudson River and the George Washington Bridge. It's cozy, though, much smaller than the grand office I first had at NYSPI when I took the administrative position as clinical director. A mid-level researcher's office. And it's in the midst of renovation, with file cabinets and boxes of books in the hallway and just a desk and a few chairs inside. The coffee-stained carpeting was just removed yesterday, and the walls are spackled, ready for repainting. Outside, a tugboat fighting the tide pushes a barge up the Hudson, and behind it, near the Palisades, two small boats speed downstream.

"So here's the situation," Sander tells me. "I've been trying to recruit for a clinical trial. We've been running for over a year and only have three subjects. It's kind of a complicated situation. People suggested that I meet with you, to see if you can help."

"Sure, glad to," I say. "What are you studying?"

Sander has horn-rimmed glasses, tousled brown hair, sideburns that extend below his earlobes, and an infectious passion for research. He's Dutch, his family was Holocaust survivors, and he once told me that back in the 1800s they changed the spelling of their last name to distance themselves from their notorious relative

Karl Marx, adding an extra letter: *Markx*. Years ago, I supervised him doing evaluations in the resident clinic. Now he's an independent scientist at the institute, a faculty member with his own lab in the department's basic research core.

He asks whether I know what's going on in cancer research. Precision oncology, it's called, and it's fascinating. Oncologists are now able to identify genes causing a patient's tumor, he tells me, and once they know the exact mutation that created the cancer, they can develop personalized, or "precision," chemotherapy drugs, unique to that person.

Those advances inspired him to start similar work in neuroscience, with psychiatric and neurological patients. *Precision psychiatry*, if you will. His goal is to find the specific genes responsible for brain diseases, whether schizophrenia or seizure disorder. He focuses on people from isolated, inbred populations, whose illness often arises from duplication of certain DNA segments. He hopes to make drugs to fix each person's brain—medications just as individualized as precision cancer chemo.

He tells me more about his patients. "They're Mennonite farmers, originally from the Netherlands. They live in Pennsylvania. They've been geographically isolated for generations, they originate from small founder populations. No one comes into their community, just sometimes people leave. They also have huge families. Even the ones who have genetic anomalies tend to have kids, they feel a special responsibility for everyone to procreate. They keep family trees for multiple generations. So when we identify an anomaly, it's easy to get both an affected family member *and* an unaffected family member, to bring them in to get cells and PET scans and all the other tests."

He explains the process, how in his lab they take a swab from the patient's or the healthy family member's cheek and manipulate the oral mucosal cells into becoming stem cells.

"You grow different kinds of neurons? Like in a Petri dish?"

"No, floating in solution. The cool thing," adds Sander, "is that if you give the right nutrients, they start forming into organoids, little *brains*. They develop a cortex, an amygdala, a hippocampus, and after a while you have hundreds of little brains floating in your lab." The image is mind-boggling, like something from *The Matrix Reloaded*. Once Sander grows minibrains, he compares the ones from healthy family members to minibrains grown from patients' cells. He can thus identify the exact gene mutation that underlies the neurological or psychiatric illness.

"And then what?"

He explains, gesturing widely. "We can test *hundreds* of drugs on these minibrains and find which drug specifically reverses the problem from each mutation." Eventually they hope to give that drug to the affected family member. And maybe actually *cure* their illness, whether schizophrenia or epilepsy. Even more mind-boggling.

"Why did you have so much difficulty getting people into your study?"

It's complicated, as Sander explains, having to do with the referring clinics in Pennsylvania towns, and National Institute of Health politics, and a review of chart documentation. The bottom line is that in order to get his study done, Sander needs more subjects, and he was advised to get a supervisor with expertise in running clinical trials. He remembered our discussions from years ago, about my medication studies combined with neuroimaging, how I got them done despite all sorts of obstacles. Which is why he's talking to me today. And to say the least, I'm happy to help.

"So, what are *you* up to these days?" asks Sander. "I hear you have a whole new thing going on, a second wind . . . a new chapter, is that so?"

"So it's weird," I start. And tell him my story.

But first, an aside, a digression, a flashback if you will. It will all make sense, I promise.

HORIZONS

"Cool, you're doing that study at your place?" He's thin, in black jeans and hoodie, his girlfriend next to him punching something into her iPhone. He's a medicinal chemist, he says. This conference, the annual Horizons psychedelic convention, is an amazing meeting, and he's been coming for years.

It's Friday night. We're at a preconference party, in Judson Memorial Church in the Village, and though the invitation promised dinner, there's only the most limited hors d'oeuvres. Tortilla chips, some veggie stuff on pita slices. I'm starving. We both grab flimsy plastic cups of white wine and juggle them with the paper plates. His partner stands close so we can hear over the beat. She calls herself (if I hear right) a pancosmic therapist; she wears a flowing multicolored gown. He had treatment-resistant depression, she tells me, and had been taking pills for many years. *Every* medicine. Then he tried IV ketamine, which didn't do much.

"But ayahuasca!" he breaks in, "Totally different story!" A connection through his therapist led to a weekend upstate. He took three of the four doses, he retched into a bucket, then began having visions of traveling through his own brain, watching electricity leaving the aberrant circuits that had misruled his life. He's been free of depression since then, *antidepressant* free, first time in eighteen years. "So, it's cool that you're studying these things, but you should *also* do ayahuasca." It's not clear whether he means try it or study it. Perhaps both.

Standing in line early the next morning, a crisp October day, waiting to get into the Twelfth Annual Horizons Conference at the historic Cooper Union building at Astor Place in lower Manhattan, I turn around, and the chemist and pancosmic therapist are standing right behind me. It may be eight a.m., but the line wraps around the block. Tall thin hipsters, middle-aged goateed therapist types in off-the-rack blazers, guys in leather vests with elaborately inked sleeves and silver bracelets and dreadlocks, a lanky clutch of Wall St. suits. It is fair to say I've never been to a conference like this before. I gesture to my new friends. I say, "Wow!"

"We never miss it," says the pancosmic therapist.

Badged up, checked in, I wander into the auditorium, where I am instantly transported to California circa 1974.

I honestly didn't need psychedelics then, that year when I dropped out of college and headed to San Francisco; California itself was a cosmic high.

It was the weirdest thing: I had left Palo Alto where I was staying with my sister and hitchhiked down to Santa Cruz, where I sauntered through redwood forests and crashed in some barista's living room, then I hit the road early in the morning, heading south on Highway 1. My thumb snagged drivers who took me along an amazing coastline, ending at a place called the Esalen Institute, where my ride, a psychologist from Santa Rosa, was attending a retreat.

He left me standing on the highway, my backpack weighing heavy, undecided what to do next. I saw a strange sight: a person on a blue girl's three-speed bike coming slowly up the hill. I couldn't tell if it was a guy or a girl. Once the person approached, they nodded and asked if I was going to the wedding. What wedding? I asked. I still couldn't tell. Here, they said, pointing down the hill. You're welcome to come. Oh, okay, I said, and I followed, but instead of going down the driveway, they wheeled the bike into the brush and started bushwhacking across the hillside. A few yards in, they shoved it under a bush, told me to stow my backpack—which I doubted I'd ever see again—and we advanced through dense shrubbery toward bright light.

We emerged onto an emerald lawn overlooking the Pacific Ocean. People were lying on the grass, or walking around in white robes, or swimming naked in a shimmering pool. The wedding clearly was over, but there were plastic cups of white wine, which we helped ourselves to. There was no food; I was famished. *(Maybe that's why I flash back from the Horizons conference so many years later: the same hungry buzz?)* The next several hours, we wandered around the grounds. No one paid any mind to us in our sweaty T-shirts and dusty jeans.

Yeah, *everyone's* tripping, he said. He was a guy, I finally concluded, a strange little guy, not a skinny girl. He seemed harmless. *You interested? I got acid.* He was biking to Mexico, taking his time. He extolled the virtues of his dietary practice, something called Dr. E. Q. Ervin's mucus-free diet. When it got late, we walked along a cliff-side path and took our clothes off and immersed ourselves in neck-deep hot sulfuric mineral baths (yeah, definitely a guy!), to watch the sun set. Then we got dressed and walked up the hillside to where we could watch the sunset a second time. *Who needs drugs?* I was thinking. He was going to crash here. He'd find a place to sleep even if it meant the bushes—did I want to stay? Instead, I retrieved my backpack from the bushes, climbed the cliff to the highway, and stuck my thumb out. I watched the sunset yet again, a third time.

Seeming hours later a pickup truck stopped for me: three Mexican laborers in the cab, no room for me. I was welcome to lie in the bed. It smelled of dirt back there, dirt and leaves and manure, and the wind whipped off the ocean. We wove down Highway 1 in the dark, and eventually they let me off at the side of the road in a little town, where I found a cheap motel. I tried to assuage my hunger by scavenging food from vending machines and sank into a dark sleep. The next morning, I hitched a few miles to Xanadu, or, more precisely, the Hearst Castle, where the bleached California hillsides were covered by grazing zebras and antelopes, and spent a day wandering through the reconstructed detritus of European castles melded into something new, born from the fantasies of the newspaper tycoon William Randolph Hearst and his mistress, the actress Marion Davies. Next, I would head to Malibu. Yeah, who needed drugs?

Saturday at Horizons flashes me back to those days, to California in the post-flower-child seventies, in patchouli-infused memory waves.

Are you experienced? Where did you travel? Almost every speaker at Horizons references their own history of psychedelic use, even the researchers. Or: *especially* the researchers, who seem impelled to join with the ambient culture. *What are your three favorite drugs?* One scientist asks himself on the podium, introducing his lecture, then provides answers. (What are *my* favorites, I ponder. Not a hallucinogen among them. Coffee and chocolate, for sure, both reassuringly plant based, but what would be my third? Alcohol, also plant derived, does only so much for me. Does ocean kayaking count, riding on plankton?)

Even in those wild California days—I ended up riding a delivery bicycle for Speedy Messenger Service, located at Pier 7 on San Francisco's Embarcadero and living in the Castro district, attending happenings in smoky San Francisco lofts, climbing Mt. Tamalpais surrounded by dazed hippies, hiking through redwood groves, going to some barely remembered party at an abandoned silver mine where Jack London once lived—even then psychedelics weren't my thing. I was the guy in the room who took a few puffs off a lumpy joint but wasn't into tripping, who waved off mysterious pills and blotter tabs. It was enough just being there, I guess. Whatever was happening, I absorbed it through my skin.

Which I guess is what you could say about the Horizons conference, today, many years later. Maybe akin to wandering through Big Sur that blazing 1970s afternoon, about to embark on an adventure, riding in the back of the pickup truck as we veered along dark cliff-side roads, exhilarating and more than a little terrifying. This is the *next* new thing, it's clear.

Am I along for the ride?

We have a prospective donor for our Columbia psychedelic studies. He's here at Horizons and has offered to show me around. So, at a break he introduces me to everyone—researchers from University College of London, from NYU, Yale, Johns Hopkins. And the heads of foundations and psychedelic advocacy groups, many of them dressed in day-glo or primary colors, flowing with energy and enchantment. The tribal devotees swirl past. *They are so damn tall, thin, young, pierced.* I see some amazing tattoos. We have lunch with Rick Doblin, the head of the MAPS Foundation—Multidisciplinary Association of Psychedelic Studies—a legendary organization founded in the dark days of 1986, two decades after Nixon destroyed psychedelic research. He is responsible for MDMA, the amphetamine derivative commonly known as Ecstasy, now nearing FDA approval for post-traumatic stress disorder. Over lunch—pizza slices and Cokes at a windy table on Astor Place—I ask whether we might run funding through his organization. After lunch, I am introduced to a Dutch researcher who is doing LSD microdose studies in healthy subjects, looking at reaction times, creative problem solving, transcendence.

In the afternoon, a psychology professor takes the stage for a brilliant diatribe. He advocates legalization of *all* drugs, especially those for which overzealous policing has incarcerated millions of minority, most notably Black, men. Crack, *yes*. PCP, *yes*. Crystal meth, *yes, yes, yes* to everything. He tells us he snorted heroin *this very morning*. It has done him no harm; he is not addicted. Rather, it has massively unleashed his creative powers. Only 30 percent of recreational heroin users get addicted; the other two-thirds just play. There are no bad drugs, he rants, only stigmatized ones.

Next, a multimedia artist's incredible film shows how she resolved sexual traumas through psychedelic drugs. We hear from a Brazilian ayahuasca healer and from a long-haired half-Colombian half-Pakistani criminologist. An angry man in the crowd yells at panel members who are trying to get FDA approval for psychedelics: *Your companies are corrupt, you can't patent shrooms, you are thieving from indigenous communities that have used them for thousands of years. Despicable! Despicable!*

A balding man squints at my Columbia ID and hands me a business card engraved with a tribal mandala. He is a former pharma executive, he whispers, now developing new psychedelic compounds through a venture-capital-funded start-up. "Shorter acting, longer acting, with *and* without hallucinatory properties!" I search my wallet and come up with a dull Columbia-issued ivory-and-blue business card, which I smooth out and hand to him.

Clearly, in no way do I fit in here, but that is fine; this is the biggest and most mind-blowing collection of zealots, scientists, shamans, reformers, libertarians, and psychonauts I have ever encountered. I'm fine with being the oddball here, one of the few attendees with only a contact high.

A WILD RIDE

Things have been tough for those of us who do clinical trials, I tell Sander that crisp fall day, a few weeks after the Horizons conference. The National Institute of Mental Health doesn't fund clinical trials anymore, plus the big pharma companies have almost all backed out of developing new psychiatric drugs. After years of difficulty getting funded, an unprecedented drought hit the Depression Evaluation Service, accelerating the retirement plans of most of our long-time faculty. We were on the edge of closing up shop, leaving me stranded.

"Then we got a trickle of new studies," I tell Sander. Following the resurgence of ketamine use, a drug company developed esketamine, the left-handed molecule of ketamine (think scissors, which can be left- or right-handed), brand name Spratavo, which we helped test as a treatment for suicidal depression.

"I heard about the Spratavo study. How'd it go?"

"Pretty cool," I tell him. "We saw interesting effects—like, while the drug is actually being infused, the patient suddenly says, "Hey, I don't feel suicidal anymore."

"Just like that?"

"Pretty much. Sometimes while it's literally going into their veins, other times a few hours afterward. Anyhow, then we started looking at a hormone, ganaxolone, which is being developed for postpartum depression. That's from another drug company. You set up the IV and stream in the drug and within a few hours—*boom!*—they tell you the depression is gone."

"Incredible!"

The cool thing about these treatments, I tell Sander, is how rapidly they work. Rather than taking a Zoloft pill every day for months and waiting for your depression to improve, these new treatments seem to blast brain circuits, changing their activity in a matter of minutes, possibly even *resetting* the circuits in some way. Many of the common psychiatric disorders seem to involve overactive, hyperconnected circuits that have gotten out of control, whether fear circuits, habit circuits, or reward circuits.

"So that's a cool question: can psychiatry find the best ways to reset them quickly?"

"We've started looking at a bunch of other drugs too," I add. "But then out of the blue, we have a new mission." I explain that we have started studying psilocybin for treatment-resistant depression. Psilocybin, originally derived from "magic mushrooms," from a wide variety of gilled mushrooms in the family Hymenogastraceae, is now synthesized in a lab. After decades of shutdown starting in the 1970s with President Nixon's War on Drugs, research on psychedelic drugs is starting to make a comeback. Brave scientists have kept the flame alive since then, along with underground communities of spiritual seekers and psychonauts who have kept using LSD, psilocybin mushrooms, and ayahuasca and tracking effects of their trips, guided by gurus and networks of rogue therapists and inspired and sometimes wacky healers. Now there's a worldwide groundswell of advocates, thousands of them, pushing for resumption of studies.

"Like Michael Pollan says in *How to Change Your Mind*."

"Exactly. Great book!" Like other new treatments, psychedelic drugs also seem to have a rapid effect on those disordered brain circuits, potentially resetting the brain. Except they aren't exactly new drugs: they are *old* drugs, many of them ancient compounds that have been used for thousands of years in religious and spiritual rituals and that are now being repurposed to treat psychiatric disorders. "Anyhow, the Food and Drug Administration has fast-tracked psilocybin for depression. And MDMA, Ecstasy, for treatment of post-traumatic stress disorder."

Consequently, I tell him, we've started up our local site, one of twenty-odd worldwide, for the Phase 2B study from Compass Pathways, using a synthetic form of psilocybin, the active chemical in magic mushrooms. Plus, I'm in the midst of planning half a dozen other psychedelic studies. Anxiety disorders, eating disorders, suicidality. And a psilocybin microdose study, which would be the first double-blind study in the world using tiny doses of psychedelics to treat patients with a psychiatric disorder.

"Amazing!"

"Yeah, it's been a wild ride."

It's taken us a year to get the psilocybin depression study underway, first negotiating the contracts and budgets; then getting approval from NYSPI's Institutional Review Board, or IRB, which reviews studies to make sure they are conducted in a safe and ethical way; followed by months of training therapists to guide the patients through their psychedelic experience.

The most daunting step, since psilocybin is a Drug Enforcement Administration (DEA) Schedule 1 drug—defined as a drug with no approved medical uses and a high potential for abuse—was getting my own Schedule 1 license to prescribe psilocybin, which involved endless background checks and reviews by the FDA and DEA, then having a New York State Bureau of Narcotics Enforcement officer, a stocky man in a day-glo tangerine three-piece suit, come for a visit to ensure secure storage in the pharmacy.

"My wife and I had a plan for downsizing and slowing down, for selling our house now that the kids are launched, and starting a simpler life. We did sell the house, we moved into the city. We were *so* ready for our lives to get a lot easier."

"So it didn't work out that way?"

I laugh. "Nope, just the opposite." Suffice it to say, other than moving to the city, things have rapidly gone in the other direction. Already, only a few weeks after the Horizons conference, I'm corresponding with researchers all over the world, talking to venture capitalists, medicinal chemists, gurus, and spiritual guides. It's a whirlwind.

"The pharmacy, it's like *Oceans 11*," I tell Sander. "The psilocybin is stored in a lockbox bolted to the wall in the research pharmacy. Inside this huge walk-in bank vault. The lockbox has two keys, one for me and one for the pharmacist. We need to turn the keys simultaneously to get to the study drug. The only things we don't have are the lasers and the somersaulting acrobats."

He laughs. "You have trouble getting patients too?"

"Hardly. From day one, we've been fending off nonstop phone calls and hundreds of emails from people wanting to be in our study."

"Sounds great. So far, how's it going?"

"We're just about to treat our first subject. Next Wednesday."

"Cool," he says. "Hey, I've got to go. See you next week?"

We plan to meet at his lab, in the Kolb Building, across Riverside Drive, so he can show me the floating brains.

THE COLD LIGHT OF DAY

Here's the question that any good clinical trialist needs to ask: Is there, scientifically speaking, anything to these drugs? They have the most fervid press these days. The level of expectation is off the charts. But are they perhaps just spectacular placebos? The fact that most of the new studies have been done by brave devotees and proselytizers of psychedelic drugs both increases the public's enthusiasm for their results and should give us pause.

Minuscule doses of these compounds, whether sourced in a rainforest or synthesized in a pharmaceutical laboratory, do commonly induce world-shaking experiences, often welcome, sometimes terrifying. Undoubtedly true. But they also (if you believe the published papers, not to mention the many breathless first-person accounts) can lead to relief of terrible symptoms of depression, trauma, obsessive-compulsive disorder, and addiction. Beyond that, they often open the user's eyes to a new view of the world, inducing a profound sense of interconnectedness with nature, of greater empathy with others, a sense of spiritual calm. Which may be the very way that they work for relieving suffering. If they truly work. All of which leads philosophers to innumerable existential mysteries and presents scientists with a host of practical challenges.

Are these true medication effects or merely amplified placebos? Is the drug the cause of change or the commonly described mystical experience? I begin to read about the long history of psychedelic drugs, many of which were used in religious and spiritual settings for centuries. In 1938, the Swiss chemist Albert Hofmann synthesized LSD while working for the Sandoz pharmaceutical company, and he accidentally discovered its hallucinogenic properties a few years later. Beginning in the 1950s, psychiatrists across the United States gave LSD to thousands of people as treatments for alcoholism, depression, and for anxiety in people with advanced cancer, and at the same time, the CIA was surreptitiously investigating it as a tool for war. Dozens of clinical studies were published, showing the benefits of psychedelic drugs. But following the 1970 passage of the Controlled Substances Act, President Richard Nixon's controversial legislation that initiated the War on Drugs, LSD, psilocybin, and other psychedelics were categorized as Schedule 1 drugs, with "no currently accepted medical use and a high potential for abuse." Soon all psychedelic research ground to a halt.

In the late 1970s, during my years at Stanford Medical School in California, there was little education on this topic, even though Palo Alto had recently been a hotbed of such research. With the chaos of the late 1960s and early 1970s, with the Vietnam War and the Chicago Seven and all that went with that, this first wave of psychedelic science was suppressed and became essentially a buried memory for both the general public and medical researchers. It was only thirty years later, in the early 2000s, that a psychedelic renaissance began, following publication of a Swiss study of LSD in patients with anxiety with life-threatening illness. This was followed by a Johns Hopkins psilocybin study of fifty-one participants suffering from cancer-related anxiety or depression, most of whom achieved relief of symptoms lasting six months or more. With more modern study design, this new generation of researchers confirmed earlier studies, finding not only that psilocybin decreased their subjects' depressed mood and anxiety about death in four-fifths of participants; it also improved their quality of life and renewed their sense of meaningfulness and optimism.

Notably, two-thirds of participants said their psilocybin treatment was one of the top five meaningful experiences in their lives. Follow-up studies, including ones by my new friend Stephen Ross's NYU study of psilocybin in cancer-related existential anxiety, confirmed the powerful—and life-changing—effects of psilocybin. In late 2018, when I do a review of the scientific literature for a lecture to the Columbia psychiatry residents, I realize with a shock that, although thousands of patients were studied in the first wave of psychedelic research in the 1950s and 1960s, even including that time only about one hundred patients have ever been studied in randomized clinical trials comparing psychedelic drugs to placebo.

If you read the studies closely, as I now begin to do—as a clinical trialist, a skeptical psychiatric researcher—you are inevitably left with innumerable questions. The sample sizes are so small, the investigators so enthusiastic. There is practically no way to "blind" the studies, since it is generally obvious to staff and participants whether a trip is underway. And the whole idea of "set and setting"—first popularized by Timothy Leary in the early 1960s during his respectable Harvard psychology professor days—by priming patients for their trips, maximizes everyone's expectations. There is no doubt that trips are often intense, whether thrilling or terrifying, and that they often spur phantasmagorical explorations of consciousness and awareness.

Psychedelic experiences are undoubtedly real. But are psychedelic *treatments* effective? And will they be *good* medications? Will they be safe enough to give to people with complicated psychiatric conditions? Whom will they help; whom will they hurt? If they loosen brain circuits, can they make some people's circuits *too* loose? Are they effective in treating symptoms of depression, OCD, eating disorders, addiction? How many milligrams are needed? How many psychedelic doses? How long will effects of a single treatment last? If given as part of combined psychotherapy and medication treatments, what's the best type of psychotherapy to pair psychedelics with? And, at the end of the day, will they be game changers or just another tool in the toolbox?

That's why, I realize with utter immodesty, the psychedelic field now needs people like me—clinical trialists, dispassionate but rigorous investigators—to run a new generation of studies to put these expectations to the test. In psychiatry we've been a dying breed, shutting down our operations, taking retirement, focusing on our private patients, but suddenly we're needed now more than ever.

And by utter coincidence, I have near-perfect credentials: I've done a ton of clinical trials in depression, I've run studies of medicines combined with psychotherapies (and *all* psychedelic treatments are medications combined with psychotherapy), I've worked with industry, I've successfully incorporated neuroimaging into double-blind clinical trials: all these things could come in handy with a new generation of psychedelic studies. Plus, I've been at loose ends, seeking a mission in this new era of neuroscience-based psychiatry. What an amazing opportunity for a turnaround; it's (to coin a term) entirely mind blowing.

Am I in? It's utterly overwhelming, just the idea of retooling for psychedelic research. Especially since I've never studied anything that deliberately induces mystical experiences. And once I do decide to take on the challenge, I quickly realize that I will need to turn on a dime, to develop a team expert enough to run such studies, and quickly get the necessary certifications, credentials, licensing, facilities, space, and equipment.

Yet how could I refuse? I sign up for the study, psilocybin in treatment-resistant major depression, as site principal investigator for an international Phase 2B study run by Compass Pathways, a groundbreaking, unprecedented study intended to fast-track psilocybin toward FDA approval as a legal drug. In a matter of weeks, my life is taken over by the imperatives of our new study. My team coalesces, newly hired research assistants, psychologists getting training in psychedelic psychotherapy, psychiatrists training to evaluate subjects with treatment-resistant depression. Soon, I am doing little else.

Our goal is to enroll as many as thirty subjects at our site. Fair enough, full speed ahead. But this is just the beginning, as soon becomes clear. If I am going to enter the psychedelic arena, the depression study should be only the first of many investigations. And this is where things take off. There are so many issues to consider, and I'm constantly texting and emailing collaborators with new thoughts: *Blinding, dosing, neuroimaging, animal models, mechanism of action is it 5HT2A? . . . priorities of disorders to study . . . how can you blind studies when everyone knows whether the patient is tripping? If using active comparator, is it better to use amphetamine or niacin? . . . Set and Setting are key to psychedelic experiences, establishing a frame for the mind-bending experiences, but how to optimize for treatment of psychiatric disorders? What kind of psychotherapy should you combine with the psychedelic drug? . . . How many doses to provide? And which drug is best for which disorder—LSD, psilocybin, DMT, MDMA? What dose works best? How many doses should you give the man with PTSD, the woman with bulimia? How long will they stay well? Are booster treatments needed? What about microdoses, are they worth studying?*

All these questions whirl in my head, they wake me after midnight, and I grab my phone to send shorthand texts about a whole new world.

THE AMAZING THING

Then there's the neuroscience thing, what perhaps fascinates all my New Neuropsychiatry colleagues, the host of neuroscience watchers, most about psychedelics: how they scramble brain networks, breaking down old connections and potentially allowing a host of new inputs.

It all comes back to serotonin, which I first learned about in my long-ago Stanford Medical School days. Psilocybin, whether originating in mushrooms or medical labs, is metabolized in the body to psilocin, its active ingredient. Psilocin's structure is tantalizingly close to that of the neurotransmitter serotonin (5HT), which plays a key role in depression and anxiety disorders (varying only by a hydroxyl group being in a different position on the carbon ring and by two methyl groups on the nitrogen side-chain).

By so closely mimicking serotonin, psilocin must thoroughly confuse our brain's 5HT2A receptors. All our lives these receptors have been bombarded by one serotonin molecule after another—and now they encounter a subtly different chemical. As psilocin binds to 5HT2A in the prefrontal cortex, it activates these receptors; it stimulates them; it is an "agonist." Wild alarms sound all over the brain, innumerable sensory and cognitive and emotional circuits start clanging and bursting and shooting and exploding, a neurosensory July Fourth extravaganza.

And afterward, long after the psilocin has been metabolized and has entirely disappeared from the body, its effects linger. This is intriguing to neuroscientists and practicing doctors. Studies suggest that the brain becomes newly plastic, mutable. The British researcher Robin Carhart-Harris posits that the psychedelics introduce "entropy" into the brain. For people with disorders, perhaps their overconnected and endlessly clanging circuits can be disrupted and potentially remodeled.

Disorder, I realize anew, means "*dis*-order"—a diseased order, a dysfunctional order, as it were, but an order nevertheless, an abnormal set of instructions that has been burned into the brain's connections, its activity, and even its anatomy. After dosing, it seems, a host of newly active circuits begin to connect anew.

Psychedelics may indeed perform a global reboot, offering the possibility of obtaining a new, and hopefully more adaptive, order.

To me and, I assume, to many fellow psychopharmacologists, these drugs, as we look at them anew, are a formidable challenge—and an endlessly engaging enigma. So much exciting work to do; we're at the very beginning of a many decades' odyssey. And back at the Psychiatric Institute, suddenly everyone wants to talk to me. *Our* circuits are reorganizing, you might say, our scientific circuits; we've been given a boost of entropy and are quickly creating new "network connectivity."

My email inbox quickly fills; my voicemail reaches its capacity. Somehow the word has gotten out—however prematurely—that I'm the local psychedelic maven. I hear from researchers specializing in anxiety disorders, eating disorders, suicide. The organic chemists Dali Sames and Mike Cunningham invite me to visit their lab at the main Columbia campus at 116th Street and Broadway. Principals from venture capital firms and private equity startups email me confidentiality agreements to sign. Board members of foundations dedicated to the advancement of psychedelic treatments and drug decriminalization and legalization start sending Zoom invites. Journalists from everywhere, wanting comments on deadline.

Emails pour in from graduate students, from medical students looking for senior electives, from psychiatry residents, from oncologists working with late-stage cancer patients, and college grads looking for research assistant jobs before going for doctoral degrees. And I make a new friend, Elias Dakwar, a psychiatrist researcher in the Columbia substance abuse division who is doing studies on ketamine-assisted psychotherapy and yearns to begin studies of the classic psychedelic drugs. Not to mention calls and emails from random members of the public wanting to pick my brain or share their ideas. Dozens of strangers for whom no conventional antidepressant has worked reach out, some looking to make connections to underground psychedelic therapists (sorry, no go, it's illegal even to *refer* to these still-illicit enterprises), most desperately seeking help.

THE IMMORTAL SELF

One day there's a knock on my door, and a recent graduate of our psychiatry residency training program, Alison Hanson, pokes her head in. Ali is a new MD-PhD, endlessly enthusiastic, and she tells me that she is working as a research fellow in the laboratory of the eminent Columbia neuroscientist Rafael Yuste, doing studies on Hydra.

"Hydra, what's that?"

"You know those tiny invertebrates, phylum Cnardia, *Hydra vulgaris*, they live in fresh water, they never age, they're immortal?"

"Yes, I think I recall from high school biology class long ago. They have lots of tentacles, they can regenerate. . . . But can you tell me how're they relevant to psychedelics?"

"You've done research on the Default Mode Network, the DMN, right?"

I nod.

"Well, Hydra are the first organism to have evolved a DMN. They're the first organism with a self."

"A self? 'Look, Mom, I'm waving my tentacles . . .' "

"Not the way *we* think of a self, but a self as an organizing network for neuronal function," Ali says. "And Hydra are the simplest creature with such a self. We can do psychedelic research on them, see how these drugs affect their DMNs. Right? We all want to know how psychedelics can affect the human self? Well, I have an animal model."

"That's wild," I say, only a bit skeptical. "I can't wait to see what you find!"

Soon I am building castles in the air, designing a host of new studies, writing protocols and grant proposals on eating disorders, OCD, and suicidality, creating the scaffolding for at least a decade of new investigations. Ali Hanson writes a grant application for studying psychedelics in *Hydra vulgaris*, naming me as a mentor. I get involved with my new friend Elias Dakwar, along with our departmental chairman, writing a 14-million-dollar proposal to set up a psychedelic center at Columbia.

Joanna Steinglass, a perspicacious young doctor and eating disorders researcher, works with me and other colleagues on a proposal for a study of psilocybin treatment of anorexia and bulimia, two eating disorders that are extremely difficult to treat.

"I get happy thinking about these drugs," Joanna tells me. "Not taking them, but studying them, they're so interesting!"

BURNING BRIGHT

Is this worth mentioning? I have the dream again, early Wednesday morning, just a few hours before our first psilocybin dosing. The one I've had since childhood, though this time not exactly the same.

I am playing ping-pong with my childhood friend Paul in an unfamiliar room—a garage, a sunny warehouse—and young tigers are wandering around us. It is very vivid, frighteningly so, with the tigers, beautiful young animals with sharp teeth and bright orange phosphorescent stripes. It's like tripping, almost. I waken, and the image stays with me as I head for work, taking my bike out of the basement and pedaling over toward Riverside Park. My new routine, biking to work: enhancing my own neurotrophic factors, not to mention cardiovascular health.

I ride along the park, then head down a long slope on the road that will lead me to the riverside path. It's a breezy morning, and most of the bikes are heading downtown. The Hudson is bright, flickering with morning light, and in the distance, far ahead, I see the high towers of the George Washington Bridge.

I start to wonder about things as I ride. What should we make of this stage, the odd resurgence of the psychedelics, this flashback to a world half a century in the past? Is it merely a throwback to an earlier era? Does it make the new neuroscience irrelevant? I don't think so. Strangely, I conclude, the psychedelics may pull everything together.

The importance of the unconscious mind in treatment from the Age of the Couch becomes strikingly relevant now. Psychedelic treatments seem to tear open the unconscious mind, make it phantasmagorically apparent in our waking life, perhaps dazzling our patient out of their suffering. Do the psychedelics invalidate the *DSM-5*? Not really, I muse, perhaps not at all. After all, they might alleviate the terrible symptoms of these disorders with only a few doses, the negative thoughts, the agitation and hopelessness and fear, quickly restoring normal mood and thought patterns. So they could be useful treatments for *DSM*-defined disorders.

What do they tell us about neuroscience? Fascinatingly, they may turn out to be one of our most powerful neuroscience tools, with the capability of radically reshaping brain circuits that have gotten stuck in dysfunctional closed loops, of introducing healthy entropy and inducing new plasticity in brain connections, maybe even restoring health to neurons. And helping us understand what regulates the brain's circuits, both in illness and recovery.

And for me, with my experience in clinical trials, there is so much work to do. Given the hype in the general media, the proselytizing from charismatic gurus and evangelical researchers alike, we expect huge placebo effects; we need to study these drugs in the context of unrealistic hopes. We need to find out what their true risks are with patients, people with severe depression and trauma histories, who need to stop their other medicines in order to get placebo treatment. We truly don't know who will benefit from these drugs and who will be hurt. Or how to best use them in treating large populations, especially once they are released to general use with broader groups of patients than those entered in studies, with less oversight of treatment centers. There's a huge opportunity here and also a huge risk, even of a second prohibition if they aren't studied well. We need to do top-flight clinical research studies.

A few miles uptown, I turn up a ramp that winds under the West Side Highway, then pedal up the long green-striped hill that leads toward the medical center. It's strange how these drugs are opening new doors for me at this time of life, resolving my latest life crisis, opening the doors of *my* perceptions, both medical-psychiatrically and literarily, to paraphrase Aldous Huxley.

I wind along the streets of Washington Heights, through the NYPresbyterian Hospital campus, heading toward the basement garage of the institute, where the bike racks are. That's when the phosphorescent tigers come back to me. These times are indeed worthy of a phosphorescent dream in which lithe orange cats stalk around the anteroom of my mind. What a strange dream! Tigers close by as we hit the feather-light ball with our ping-pong paddles. And even half-asleep I realized that this dream connected to my recurrent nightmare from childhood of a tiger at the screen door of my family's home in Cleveland Heights.

Last night's dream was different from the one that terrified me long ago, I realize as I click the Kryptonite lock shut and head toward the elevators that lead to NYSPI's lobby. The young tigers, so phosphorescent, almost *fluorescent*, were as brilliant as optogenetically labeled genes!

Somehow the image makes me laugh. It must come from meeting Sander the other day, with our talk of new studies. With all the young researchers at the institute, MDs and PhDs, their expertise in optogenetics and PET imaging and "big data," skulking around us old-fashioned *DSM* psychiatrists who are trying to become late-life neuroscientists. Carnivorous predators? Or, perhaps, as young cats stretch in the sunlight, they're ready to play.

DOSING

A few minutes later, still sweaty, I stand in front of the New York State Psychiatric Institute Research Pharmacy door. I knock a bunch of times before a drowsy pharmacy assistant lets me in. Robert, the head pharmacist, then appears. He unlocks the main walk-in vault, turns his key in the steel-bolted lockbox, and pops back out. I lunge into the vault, the door clanging behind me, and turn my MD key in the lockbox.

Three shelves hold identical white bottles of study drug.

My iPhone confirms the bottle code assigned to Subject 212-001. I make sure to grab the correct vial, to complete the dispensation log: bottle number, lot number, expiration date.

4331
22459B
02/NOV/2019

Upstairs I follow the block-long curved hallway to the dosing room in the Biological Studies Unit at the south wing. The fluorescents are dark because psilocybin can increase light sensitivity; electric candles flicker ahead. A wooden sign reads:

STUDY IN PROGRESS
DO NOT DISTURB

Inside the dosing room, lights are dimmed, and curtains cover the windows. Our patient, a thirty-something guy who has been depressed since his early teens, reclines on the bed, nervous but psyched. His hands shake as he rips plastic off disposable eyeshades. Giant noise-cancelling headphones sit on his lap and will soon stream a custom musical playlist that will ease him through his trip. Our two clinical psychologists, Elizabeth and Martin, enter and get comfortable in their recliners, since they'll be with him from start to finish.

"So, are we ready?"

Our patient sits forward on the bed. I inspect the vial again: bottle number, lot number, expiration date. "Give me your hand," I say. I shake out the five white capsules into his palm.

"We really should get a chalice," Elizabeth says. "That's what they use in some sites."

He's holding a plastic cup of water. He places the five soft white capsules in his mouth, and we watch him wash them down.

"Everyone okay?"

"We're good."

"I'll check on you at the end of the day," I tell him. "Your therapists will let me know how things are going." He nods. "Have a good trip."

He leans on the bolster, adjusts his eyeshades and the noise-canceling headphones. Elizabeth checks the sound level, then covers his legs with a blanket.

I close the door. A thought pops into my mind: What if Sander gave psilocybin to those floating minibrains? Those little self-organizing organoids? What would they do? What effects on cell structure, architecture, network connectivity? Sounds kind of nutty, but I make a note to ask when we meet next week in his lab. Maybe he has a postdoc with free time. Somehow nothing is nutty these days.

Back in my office, I start unpacking boxes. I can smell the glue under new linoleum, and barely dry paint. A while later, I get a text from one of the therapists:

> Hi David—
>
> An hour in and all proceeding smoothly.
>
> Martin

So—it's starting, this phantasmagorical new adventure. Who can imagine what comes next?

Afterword

Parts of this book, precursors really, first appeared in magazines, in the forms of essays and articles in places as diverse as the *New York Times* health and science sections, the *New York Times Magazine*, the literary magazine *North American Review*, and in journals with medical audiences such as *Health Affairs*, *Hippocrates*, *Massachusetts Medicine*, and Columbia University Medical School's literary journal *Reflexions*. More recently they have appeared online in sites including *Psychologytoday.com* and *Huffingtonpost.com*. In preparing them for publication in a single volume, it became clear that there was often a broader underlying theme beyond the immediate essay or article, which upon further exploration could better illuminate the ever-evolving world of psychiatric practice, including the jarring transitions between epochs, the "disruptive" changes as new technologies emerge and as old ones struggle against annihilation.

As a result, I have reworked and transformed most of these essays. For instance, "The Work," which originally appeared in *Living Well* magazine, has been expanded to evoke the intense, at times suffocating, culture of psychoanalytic psychiatry when I trained at Manhattan's old Payne Whitney Clinic. The character of Ms. J is an invention, consolidating vivid memories of the struggles to learn psychodynamic psychotherapy, in which dilemmas, conflicts, and other emotional memories are often more vivid than the specifics of setting or person. "Flights Into Health," an abbreviated version of which first appeared in the daily *New York Times*, goes beyond its original limits to talk about one of the most amazing and controversial new developments in neuroscience research: that the human brain continues to develop new neurons throughout life, a "neurogenesis" crucial to the effectiveness of antidepressant medicines. I have added other essays as well to fill in gaps—in my career and life, and even more, in the practice of psychiatry, especially as the psychedelic drugs have begun to undergo an unprecedented revival on the cutting edge of treatment and science. Throughout, to preserve the privacy of patients I have worked with in clinical and research settings, I have altered identifying information, names, genders, ages, and circumstances of treatment. All cases described grow from a nub of real experience: a dilemma, an impasse, a conflict, a breakthrough, an unexpected trajectory. Often remembered decades later, the emotional, medical, and psychiatric realities were far more vivid than the wispy details of biography. Thus, I have elaborated from a firm base of reality throughout *The Couch, the Clinic, and the Scanner,* not only with Ms. J, but also with Hank, Viv, Maureen, and others, creating essentially fictional characters with full lives, in the hope of revealing truth and exploring mysteries, and telling honest tales. This book in a sense is a work of metafiction, or perhaps metanarrative, in that the teller (and perhaps the reader) has been changed by the telling of the tales.

I am grateful for readers of previous versions of this book, including Samuel Douglas, Nat Sobel, Judith Weber, and Sara Paolozzi, whose thoughtful and sympathetic readings helped me make useful revisions, and to the late Daniel Menaker, a former *New Yorker* editor, who advised me on refocusing my concept of the collection, and to Jenny Davidson who advised on publication. I am especially grateful to Miranda Martin and her colleagues at Columbia University Press, who have guided the manuscript through the final phases of revision.

I would particularly like to thank my wife, Lisa Perry Hellerstein, for her infinite patience and loving support over the decades; our three inquisitive, creative, and challenging children, Sarah, Benjamin, and Jason; and my various editors over the years, particularly Robley Wilson of the *North American Review,* who helped me find my voice as a writer, starting from my first efforts at essay writing when I was starting residency training at New York Hospital/Payne Whitney Clinic. I also owe a debt of gratitude to the MacDowell Colony and the Virginia Center for Creative Arts, the two arts colonies where I have found periods of respite during the many-year gestation of *The Couch, the Clinic, and the Scanner,* and to my colleagues at the numerous hospitals and clinics where I have worked and consulted over the years, in California, Massachusetts, New York, and elsewhere, whose work and ongoing dialogues have helped me understand the directions of psychiatry over three miraculous ages. I feel especially grateful to my patients, from whom I've learned to be a doctor and psychiatrist and to have measures of hope and humility and who have continually challenged me to learn more and be a better healer, a better doctor, a better psychiatrist, and perhaps even a better person.

References

PREFACE

Kendler, K. S., K. Tabb, and J. Wright. "The Emergence of Psychiatry, 1650–1850." *American Journal of Psychiatry* 179, no. 5 (May 2022): 329–35.

1. THE WORK: LEARNING TO DO PSYCHOANALYTIC PSYCHOTHERAPY, 1980–1984

Freud, S. "Observations on Transference-Love (Further Recommendations on the Technique of Psycho-analysis III)." In *The Standard Edition of the Complete Psychological Works of Sigmund Freud*, vol. 12, ed. James Strachey et al. London: Hogarth, 1955.

—. "Recommendations on the Technique of Psycho-Analysis." In *The Standard Edition of the Complete Psychological Works of Sigmund Freud*, vol. 12, ed. James Strachey et al. London: Hogarth, 1955.

Gabbard, G. O. "A Contemporary Psychoanalytic Model of Countertransference." *Journal of Clinical Psychology* 57, no. 8 (2001): 983–91.

Heimann, P. "On Counter-transference." *International Journal of Psycho-analysis* 31 (1950): 81–84.

Reich, A. "On Counter-transference." In *Classics in Psychoanalytic Technique*, ed. R. Langs, 153–59. Northvale, NJ: Jason Aronson, 1951.

Rifkin, A., D. F. Klein, D. Dillon, and M. Levitt. "Blockade by Imipramine or Desipramine of Panic Induced by Sodium Lactate." *American Journal of Psychiatry*, May 1981.

Saul, L. J. "The Erotic Transference." *Psychoanalytic Quarterly* 31, no. 1 (January 1962): 54–61.

2. TIGERS IN THE NIGHT: A THERAPIST'S OWN THERAPY, 1981–1988

Freud, S. *The Interpretation of Dreams*. 3rd ed. Trans. A. A. Brill. New York: Macmillan, 1913.

3. THE ENCHANTED GARDEN: PSYCHOANALYSIS IN THE PSYCHIATRY MARKETPLACE, 1985

Borges, J. L. *Ficciones*. New York: Grove, 1994.

Hellerstein D., W. Frosch, and H. W. Koenigsberg. "The Clinical Significance of Command Hallucinations." *American Journal of Psychiatry*, February 1987.

4. DREAMS OF THE INSANE HELP GREATLY IN THEIR CURE: DEMOLITION OF THE PSYCHOANALYTIC MOTHERSHIP, 1994

American Psychiatric Association. *DSM-II. Psychiatric Diagnostic & Statistical Manual of Mental Disorders*. Washington, DC: American Psychiatric Association Press, 1968.

—. *DSM-III. Psychiatric Diagnostic & Statistical Manual of Mental Disorders*. Washington, DC: American Psychiatric Association Press, 1980.

Freud, S., J. Strachey, and A. Richards. *Introductory Lectures on Psychoanalysis*. London: Penguin, 1991.

Hamilton, I. *Robert Lowell: A Biography*. London: Faber and Faber, 2011.

Hellerstein, D. "Letting Go of Payne Whitney." *New York Times Magazine*, November 6, 1994.

Jamison, K. R. *Robert Lowell: Setting the River on Fire*. New York: Random House, 2018.

Lowell, R. *Collected Poems*. New York: Farrar, Straus and Giroux, 2003.

—. *Life Studies*. New York: Farrar, Straus and Cudahy, 1959.

5. TREATING THE CITY: *DSM* PSYCHIATRY IN THE REAL WORLD OF THE CITY HOSPITAL, 1989

Cuellar, A. E., and P. J. Gertler. "Trends in Hospital Consolidation: The Formation of Local Systems." *Health Affairs* 22, no. 6 (2003): 77–87.

Gaynor, M., and R. Town. "The Impact of Hospital Consolidation—Update." The Synthesis Project, Robert Wood Johnson Foundation, June 2012. http://www.rwjf.org/content/dam/farm/reports/issue_briefs/2012/rwjf73261.

6. REINVENTING THE EGG: TRANSLATING THE *DSM* ACROSS CULTURES AND LANGUAGES, 1990–1994

Desbiens, N. A., and H. J. Vidaillet. "Discrimination Against International Medical Graduates in the United States Residency Program Selection Process." *BMC Medical Education* 10, no. 1 (2010): 5.

Hagopian, A., M. J. Thompson, E. Kaltenbach, and L. G. Hart. "Health Departments' Use of International Medical Graduates in Physician Shortage Areas." *Health Affairs* 22, no. 5 (2003): 241–49.

Hart, L. G., S. M. Skillman, M. Fordyce, et al. "International Medical Graduate Physicians in the United States: Changes Since 1981." *Health Affairs* 26, no. 4 (2007): 1159–69.

7. THE RED BOX: DIGGING DEEP INTO THE *DSM*, LATE 1990s

Groves, J. E. "Taking Care of the Hateful Patient." *New England Journal of Medicine* 298, no. 16 (April 20, 1978): 883–87.

Siskind, D., V. Siskind, and S. Kisely. "Clozapine Response Rates Among People with Treatment-Resistant Schizophrenia: Data from a Systematic Review and Meta-Analysis." *Canadian Journal of Psychiatry* 62, no. 11 (2017): 772–77.

8. CALL: TESTING THE *DSM* OFF HOURS, 1998

Asch, D. A., and R. M. Parker. "The Libby Zion Case." *New England Journal of Medicine* 318 (1988): 771–75.

Fins, J. J. "Professional Responsibility: A Perspective on the Bell Commission Reforms." *Bulletin of the New York Academy of Medicine* 67, no. 4 (1991): 359.

Holzman, I. R., and S. H. Barnett. "The Bell Commission: Ethical Implications for the Training of Physicians." *Mount Sinai Journal of Medicine* 67, no. 2 (2000): 136–39.

9. LESS WITH LESS: STRIPPING THE *DSM* TO THE ESSENTIALS OR BEYOND, 1998–2000

Baadh, A., Y. Peterkin, M. Wegener, et al. "The Relative Value Unit: History, Current Use, and Controversies." *Current Problems in Diagnostic Radiology* 45, no. 2 (2016): 128–32.

Glied, S. "Managed Care." In *Handbook of Health Economics*, 1:707–53. Elsevier, 2000.

10. FLIGHTS INTO HEALTH: LEARNED SAFETY AND THE NEW NEUROPSYCHIATRY, 2000–2007

Frodl, T. S., N. Koutsouleris, R. Bottlender, et al. "Depression-Related Variation in Brain Morphology Over 3 Years: Effects of Stress?" *Archives of General Psychiatry* 65 (2008): 1156–65.

Jovanovic, T., A. Kazama, J. Bachevalier, and M. Davis. "Impaired Safety Signal Learning May Be a biomarker of PTSD." *Neuropharmacology* 62, no. 2 (2012): 695–704.

Kong, E., F. J. Monje, J. Hirsch, and D. D. Pollak. "Learning Not to Fear: Neural Correlates of Learned Safety." *Neuropsychopharmacology* 39 (2014): 515–27.

Manganas, L. N., X. Zhang, et al. "Magnetic Resonance Spectroscopy Identifies Neural Progenitor Cells in the Live Human Brain." *Science* 318 (2007): 980–98.

Michopoulos, V., S. D. Norrholm, J. S. Stevens, et al. "Dexamethasone Facilitates Fear Extinction and Safety Discrimination in PTSD: A Placebo-Controlled, Double-Blind Study." *Psychoneuroendocrinology* 83 (2017): 65–71.

Pollak, D. D., et al. "A Translational Bridge Between Mouse and Human Models of Learned Safety." *Annals of Medicine* 42, no. 2 (2010): 127–34.

Pollak, D. D., F. J. Monje, L. Zuckerman, et al. "An Animal Model of a Behavioral Intervention for Depression." *Neuron* 60 (2008): 149–61.

Rogan, M. T., K. S. Leon, D. L. Perez, and E. R. Kandel. "Distinct Neural Signatures for Safety and Danger in the Amygdala and Striatum of the Mouse." *Neuron* 46 (2005): 309–20.

11. CURING FAMILIES: GENES, CIRCUITS, AND THE FRONTIERS OF TREATMENT, 2005–2009

Aguilera, M., B. Arias, M. Wichers, et al. "Early Adversity and 5-HTT/BDNF Genes: New Evidence of Gene-Environment Interactions on Depressive Symptoms in a General Population." *Psychological Medicine* 39, no. 9 (2009): 1425–32.

Assary, Elham, John Paul Vincent, Robert Keers, and Michael Pluess. "Gene-Environment Interaction and Psychiatric Disorders: Review and Future Directions." In *Seminars in Cell and Developmental Biology*, 77:133–43. Academic Press, 2018.

Caspi, A., K. Sugden, T. E. Moffitt, et al. "Influence of Life Stress on Depression: Moderation by a Polymorphism in the 5-HTT Gene." *Science* 301, no. 5631 (2003): 386–89.

Franklin, T. B., H. Russig, I. C. Weiss, et al. "Epigenetic Transmission of the Impact of Early Stress Across Generations." *Biological Psychiatry* 68, no. 5 (2010): 408–15.

Hellerstein, D. "One by One a Family Is Treated." *New York Times*, February 26, 2002.

Kendler, K. S., J. W. Kuhn, J. Vittum, et al. "The Interaction of Stressful Life Events and a Serotonin Transporter Polymorphism in the Prediction of Episodes of Major Depression: A Replication." *Archives of General Psychiatry* 62, no. 5 (2005): 529–35.

Yehuda, R., and L. M. Bierer. "Transgenerational Transmission of Cortisol and PTSD Risk." *Progress in Brain Research* 167 (2007): 121–35.

12. OFF LABEL: REVISIONING DRUGS IN THE AGE OF NEUROSCIENCE, 1997–2023

Addolorato, Giovanni, et al. "Baclofen Efficacy in Reducing Alcohol Craving and Intake: A Preliminary Double-Blind Randomized Controlled Study." *Alcohol and Alcoholism* 37, no. 5 (2002): 504–8.

Angell, Marcia. *The Truth About the Drug Companies: How They Deceive Us and What to Do About It*. New York: Random House, 2005.

Beal, B., T. Moeller-Bertram, J. M. Schilling, and M. S. Wallace. "Gabapentin for Once-Daily Treatment of Post-herpetic Neuralgia: A Review." *Clinical Interventions in Aging* 7 (2012): 249.

Berlin, R. K., P. M. Butler, and M. D. Perloff. "Gabapentin Therapy in Psychiatric Disorders: A Systematic Review." *Primary Care Companion for CNS Disorders* 17, no. 5 (2015).

Calandre, E. P., F. Rico-Villademoros, and M. Slim. "Alpha2delta Ligands, Gabapentin, Pregabalin, and Mirogabalin: A Review of Their Clinical Pharmacology and Therapeutic Use." *Expert Review of Neurotherapeutics* 16, no. 11 (2016): 1263–77.

Clarke, H., et al. "The Prevention of Chronic Postsurgical Pain Using Gabapentin and Pregabalin: A Combined Systematic Review and Meta-analysis." *Anesthesia & Analgesia* 115, no. 2 (2012): 428–42.

Dichtel, L. E., M. Nyer, C. Dording, et al. "Effects of Open-Label, Adjunctive Ganaxolone on Persistent Depression Despite Adequate Antidepressant Treatment in Postmenopausal Women: A Pilot Study." *Journal of Clinical Psychiatry* 81, no. 4 (2020): 7602.

Eguale, T., D. L. Buckeridge, A. Verma, et al. "Association of Off-Label Drug Use and Adverse Drug Events in an Adult Population." *JAMA Internal Medicine* 176, no. 1 (2016): 55–63.

Fava, M., S. Stahl, L. Pani, et al. "REL-1017 (Esmethadone) as Adjunctive Treatment in Patients with Major Depressive Disorder: A Phase 2a Randomized Double-Blind Trial." *American Journal of Psychiatry* 179, no. 2 (2022): 122–31.

Fawcett, J., A. J. Rush, J. Vukelich, et al. "Clinical Experience with High-Dosage Pramipexole in Patients with Treatment-Resistant Depressive Episodes in Unipolar and Bipolar Depression." *American Journal of Psychiatry* 173, no. 2 (2016): 107–11.

Green, B. "Prazosin in the Treatment of PTSD." *Journal of Psychiatric Practice* 20, no. 4 (2014): 253–59.

Houghton, K. T., A. Forrest, A. Awad, et al. "Biological Rationale and Potential Clinical Use of Gabapentin and Pregabalin in Bipolar Disorder, Insomnia, and Anxiety: Protocol for a Systematic Review and Meta-analysis." *BMJ Open* 7, no. 3 (2017): e013433.

Mason, B. J., et al. "Gabapentin Treatment for Alcohol Dependence: A Randomized Clinical Trial." *JAMA Internal Medicine* 174, no. 1 (2014): 70–77.

McElroy, S. L., et al. "A Pilot Trial of Adjunctive Gabapentin in the Treatment of Bipolar Disorder." *Annals of Clinical Psychiatry* 9, no. 2 (1997): 99–103.

Moore, R. A., P. J. Wiffen, S. Derry, and A. S. Rice. "Gabapentin for Chronic Neuropathic Pain and Fibromyalgia in Adults." *Cochrane Database of Systematic Reviews* 4 (2014).

Pande, A. C., et al. "Gabapentin in Bipolar Disorder: A Placebo-Controlled Trial of Adjunctive Therapy 1." *Bipolar Disorders* 2, no. 3p2 (2000): 249–55.

Rowbotham, M., et al. "Gabapentin for the Treatment of Postherpetic Neuralgia: A Randomized Controlled Trial." *Journal of the American Medical Association* 280, no. 21 (1998): 1837–42.

Statistica.com. "Number of Gabapentin Prescriptions in the U.S. from 2014 to 2019." https://www.statista.com/statistics/781648/gabapentin-prescriptions-number-in-the-us/.

Steinman, M. A., L. A. Bero, M. M. Chren, and C. S. Landefeld. "Narrative Review: The Promotion of Gabapentin: An Analysis of Internal Industry Documents." *Annals of Internal Medicine* 145, no. 4 (2006): 284–93.

Thielking, M. "Ketamine Gives Hope to Patients with Severe Depression. But Some Clinics Stray from the Science and Hype Its Benefits." *Statnews.com*, September 24, 2018. https://www.statnews.com/2018/09/24/ketamine-clinics-severe-depression-treatment/.

Vedula, S. S., L. Bero, R. W. Scherer, and K. Dickersin. "Outcome Reporting in Industry-Sponsored Trials of Gabapentin for Off-Label Use." *New England Journal of Medicine* 361, no. 20 (2009): 1963–71.

Wallach, J. D., and J. S. Ross. "Gabapentin Approvals, Off-Label Use, and Lessons for Postmarketing Evaluation Efforts." *Journal of the American Medical Association* 319, no. 8 (2018): 776–78.

13. MIND WANDERING, THEN AND NOW: NEW VIEWS OVER THREE ERAS, 2005–2023

Bansal, R., D. J. Hellerstein, and B. S. Peterson. "Evidence for Neuroplastic Compensation in the Cerebral Cortex of Persons with Depressive Illness." *Molecular Psychiatry* 23 (2018): 375.

Bansal, R., D. J. Hellerstein, S. Sawardekar, et al. "Effects of the Antidepressant Medication Duloxetine on Brain Metabolites in Persistent Depressive Disorder: A Randomized, Controlled Trial." *PloS One* 14, no. 7 (2019): e0219679.

Brewer, J. A., et al. "Meditation Experience Is Associated with Differences in Default Mode Network Activity and Connectivity." *Proceedings of the National Academy of Sciences* 108, no. 50 (2011): 20254–59.

Charney, D., C. Nemeroff, and S. R. Braun. *The Peace of Mind Prescription: An Authoritative Guide to Finding the Most Effective Treatment for Anxiety and Depression*. New York: Houghton Mifflin, 2004.

Feder, A., S. F. Torres, S. M. Southwick, and D. S. Charney. "The Biology of Human Resilience: Opportunities for Enhancing Resilience Across the Lifespan." *Biological Psychiatry* 86, no. 6 (2019).

Freud, S., J. Strachey, and A. Richards. *Introductory Lectures on Psychoanalysis*. London: Penguin, 1991.

Menon, V., and L. Q. Uddin. "Saliency, Switching, Attention, and Control: A Network Model of Insula Function." *Brain Structure and Function* 214, nos. 5–6 (2010): 655–67.

Posner, J., D. J. Hellerstein, I. Gat, et al. "Antidepressants Normalize the Default Mode Network in Patients with Dysthymia." *JAMA Psychiatry* 70, no. 4 (2013): 373–82.

Sheline, Y. I., D. M. Barch, J. L. Price, et al. "The Default Mode Network and Self-Referential Processes in Depression." *Proceedings of the National Academy of Sciences* 106, no. 6 (2009): 1942–47.

Southwick, S. M., H. Douglas-Palumberi, and R. H. Pietrzak. "Resilience." In *Handbook of PTSD: Science and Practice*, 2nd ed., ed. M. J. Friedman, P. A. Resick, and T. M. Keane, 590–606. New York: Guilford, 2014.

Wang, J., J. Bernanke, B. S. Peterson, et al. "The Association Between Antidepressant Treatment and Brain Connectivity Across Two Double-Blind Placebo-Controlled Clinical Trials." *Lancet Psychiatry* 393 (2019): 667–74.

Yehuda, R., and L. M. Bierer. "The Relevance of Epigenetics to PTSD: Implications for the *DSM-V*." *Journal of Traumatic Stress* 22, no. 5 (2009): 427–34.

Young, L. J., et al. "Increased Affiliative Response to Vasopressin in Mice Expressing the V1a Receptor from a Monogamous Vole." *Nature* 400, no. 6746 (1999): 766–68.

14. FLOATING BRAINS AND MAGIC MUSHROOMS: ANCIENT PSYCHEDELICS TEST THE PROGRESS OF PSYCHIATRY, 2019 TO TODAY

Amin, N. D., and S. P. Paşca. "Building Models of Brain Disorders with Three-Dimensional Organoids." *Neuron* 100, no. 2 (2018): 389–405.

Bender, D., and D. J. Hellerstein. "Assessing the Risk-Benefit Profile of Classical Psychedelics: A Clinical Review of Second-Wave Psychedelic Research." *Psychopharmacology* 239 (2022): 1907–32.

Carhart-Harris, R. L., M. Bolstridge, C. M. Day, et al. "Psilocybin with Psychological Support for Treatment-Resistant Depression: Six-Month Follow-up." *Psychopharmacology* 235, no. 2 (2018): 399–408.

Carhart-Harris, R. L., M. Bolstridge, J. Rucker, et al. "Psilocybin with Psychological Support for Treatment-Resistant Depression: An Open-Label Feasibility Study." *Lancet Psychiatry* 3, no. 7 (2016): 619–27.

Carhart-Harris, R. L., L. Roseman, M. Bolstridge, et al. "Psilocybin for Treatment-Resistant Depression: fMRI-Measured Brain Mechanisms." *Scientific Reports* 7, no. 1 (2017).

de Jong, J. O., C. Llapashtica, K. Strauss, et al. "Cortical Overgrowth in a Preclinical Forebrain Organoid Model of CNTNAP2-Associated Autism Spectrum Disorder." *bioRxiv* (2019): 739391.

Drew, L. J., G. W. Crabtree, S. Markx, et al. "The 22q11. 2 Microdeletion: Fifteen Years of Insights Into the Genetic and Neural Complexity of Psychiatric Disorders." *International Journal of Developmental Neuroscience* 29, no. 3 (2011): 259–81.

Drost, J., and H. Clevers. "Organoids in Cancer Research." *Nature Reviews Cancer* 18, no. 7 (2018): 407–18.

Griffiths, R. R., W. A. Richards, U. McCann, and R. Jesse. "Psilocybin Can Occasion Mystical-Type Experiences Having Substantial and Sustained Personal Meaning and Spiritual Significance." *Psychopharmacology* 187, no. 3 (2006): 268–83.

Hartogsohn, I. "Constructing Drug Effects: A History of Set and Setting." *Drug Science, Policy and Law* 3 (2017).

Johnson, M. W., A. Garcia-Romeu, and R. R. Griffiths. "Long-Term Follow-up of Psilocybin-Facilitated Smoking Cessation." *American Journal of Drug and Alcohol Abuse* 43, no. 1 (2017): 55–60.

Leary, T., G. H. Litwin, and R. Metzner. "Reactions to Psilocybin Administered in a Supportive Environment." *Journal of Nervous and Mental Disease* (1963).

Mithoefer, M. C., C. S. Grob, and T. D. Brewerton. "Novel Psychopharmacological Therapies for Psychiatric Disorders: Psilocybin and MDMA." *Lancet Psychiatry* 3, no. 5 (2016): 481–88.

Nutt, D. "Psychedelic Drugs—A New Era in Psychiatry?" *Dialogues in Clinical Neuroscience* 21, no. 2 (2019): 139.

Preller, K. H., P. Duerler, J. B. Burt, et al. "Psilocybin Induces Time-Dependent Changes in Global Functional Connectivity: Psi-induced Changes in Brain Connectivity." *Biological Psychiatry* (2020).

Pollan, M. *How to Change Your Mind: What the New Science of Psychedelics Teaches Us About Consciousness, Dying, Addiction, Depression, and Transcendence*. New York: Penguin, 2019.

Waldman, A. *A Really Good Day: How Microdosing Made a Mega Difference in My Mood, My Marriage, and My Life*. New York: Knopf, 2017.

Wang, H. "Modeling Neurological Diseases with Human Brain Organoids." *Frontiers in Synaptic Neuroscience* 10 (2018): 15.

Index

Ingram Content Group UK Ltd.
Milton Keynes UK
UKHW041808090723
424751UK00002BA/8/J